Optimization of CRISPR/Cas Editing Efficiency in Non-Model Organisms

A Non-Technical Simplest Approach for General Readers

Rickbed Nandi

Preface

Advances in CRISPR/Cas technology have revolutionized genome editing, offering unprecedented precision and efficiency in modifying genetic material. In our book, "*Optimization of CRISPR/Cas Editing Efficiency in Non-Model Organisms*," I explored the complexities of utilizing this powerful tool in research involving non-model organisms. From understanding the molecular machinery of the CRISPR/Cas system to overcoming barriers in delivery and enhancing homology-directed repair efficiency, each chapter offers a comprehensive exploration of key concepts and strategies. Additionally, I addressed the ethical and regulatory considerations inherent in genome editing, providing insights into the evolving regulatory landscape governing the use of CRISPR/Cas technology. Through evidence-based discussions and case studies, this book aims to guide researchers and policymakers in navigating the intricacies of CRISPR/Cas editing in non-model organisms, ultimately fostering responsible and impactful research practices.

Table of Contents

Chapter 1: Introduction to CRISPR/Cas Editing in Non-Model Organisms 7

    1.1 Overview of CRISPR/Cas Technology 7

    1.2 Importance of Non-Model Organisms in Research 10

    1.3 Challenges in CRISPR/Cas Editing in Non-Model Organisms 14

Chapter 2: Understanding CRISPR/Cas System Components 19

    2.1 CRISPR/Cas System: Molecular Machinery 19

    2.2 Components of CRISPR/Cas System 22

    2.3 Functionality of Cas Proteins 25

    2.4 Types of Guide RNAs 29

Chapter 3: Designing Guide RNAs for Efficient Targeting 33

    3.1 Principles of Guide RNA Design 33

    3.2 Tools and Software for Guide RNA Design 36

    3.3 Considerations for Off-Target Effects 39

    3.4 Strategies for Enhancing Guide RNA Specificity 42

3.5 Case Studies: Successful Guide RNA Design in Non-Model Organisms 44

Chapter 4: Delivery Methods for CRISPR/Cas Components 49

    4.1 Introduction to Delivery Systems 49

    4.2 Methods for Delivering Cas Protein 53

    4.3 Strategies for Delivering Guide RNA 57

    4.4 Combined Delivery Approaches 60

    4.5 Case Studies: Comparative Analysis of Delivery Methods in Non-Model Organisms 63

Chapter 5: Enhancing Homology-Directed Repair (HDR) Efficiency 68

    5.1 Mechanisms of HDR 68

    5.2 Factors Influencing HDR Efficiency 71

    5.3 Strategies for Enhancing HDR Efficiency 75

    5.4 Role of Small Molecules and Chemicals 77

    5.5 Case Studies: Optimizing HDR for Precise Genome Editing in Non-Model Organisms 80

Chapter 6: Overcoming Barriers to CRISPR/Cas Editing in Non-Model Organisms 84

    6.1 Species-Specific Challenges 84

6.2 Genetic and Genomic Variability 87

6.3 Cellular and Tissue Specificity 92

6.4 Environmental Factors..................................96

6.5 Emerging Strategies to Overcome Barriers ...100

Chapter 7: High-Throughput Screening for CRISPR/Cas Editing Optimization 106

7.1 Importance of High-Throughput Screening...106

7.2 Screening Assay Design 109

7.3 Automation and Data Analysis......................113

7.4 Applications in Non-Model Organisms 118

7.5 Case Studies: High-Throughput Screening Success Stories 122

Chapter 8: Bioinformatics Tools for CRISPR/Cas Optimization 126

8.1 Role of Bioinformatics in CRISPR/Cas Editing 126

8.2 Genome Analysis for Target Identification 129

8.3 Off-Target Prediction Tools...........................133

8.4 Data Integration and Analysis Platforms 137

8.5 Case Studies: Leveraging Bioinformatics for Enhanced Editing Efficiency..................................141

Chapter 9: Ethical and Regulatory Considerations in Non-Model Organism Editing 145

   9.1 Ethical Implications of Genome Editing.........145

   9.2 Regulatory Landscape for Non-Model Organisms 149

   9.3 Guidelines and Best Practices........................154

   9.4 Public Perception and Engagement 158

References..164

Chapter 1: Introduction to CRISPR/Cas Editing in Non-Model Organisms

1.1 Overview of CRISPR/Cas Technology

CRISPR/Cas technology has revolutionized genome editing, offering unprecedented precision, efficiency, and versatility across diverse organisms. Originating from a prokaryotic adaptive immune system, CRISPR/Cas has been harnessed as a powerful tool for targeted genome modifications in both model and non-model organisms.

At the core of CRISPR/Cas technology lies its programmable RNA-guided endonuclease activity, typically mediated by the Cas9 protein. The system relies on two main components: a guide RNA (gRNA) and the Cas protein. The gRNA directs the Cas protein to a specific DNA target sequence through Watson-Crick base pairing, enabling precise DNA cleavage at the desired genomic location. This process initiates cellular repair mechanisms, including non-homologous end joining (NHEJ) or homology-directed repair (HDR), facilitating gene knockout, knock-in, or modulation with high efficiency.

The simplicity and flexibility of CRISPR/Cas systems have fuelled their widespread adoption in diverse organisms, transcending the traditional limitations of genetic manipulation techniques. Notably, CRISPR/Cas technology has been successfully applied in a myriad of non-model organisms, ranging from plants and fungi to animals and microbes. Its utility in non-model organisms stems from its ability to target virtually any genomic locus with minimal prior knowledge of the target organism's genome, thus democratizing genome editing across species.

In plants, CRISPR/Cas-mediated genome editing has facilitated the generation of crops with enhanced agronomic traits, disease resistance, and nutritional quality. For instance, researchers have employed CRISPR/Cas to engineer resistance against devastating plant pathogens such as Xanthomonas oryzae pv. oryzae in rice, a staple food for millions worldwide. Similarly, in animals, CRISPR/Cas technology has enabled precise genetic modifications for applications in biomedical research, livestock breeding, and conservation efforts. Notably, the creation of genetically modified (GM) mosquitoes resistant to malaria or dengue virus transmission exemplifies the potential of CRISPR/Cas editing to combat vector-borne diseases.

In addition to its applications in agriculture and biomedicine, CRISPR/Cas technology has found utility in microbial biotechnology, environmental remediation, and synthetic biology. Microorganisms such as bacteria and yeast have been engineered using CRISPR/Cas for various biotechnological purposes, including the production of biofuels, pharmaceuticals, and bioplastics. Moreover, CRISPR-based genome editing holds promise for ecological applications, such as restoring endangered species' genetic diversity or engineering microbial communities for environmental cleanup.

Despite its transformative potential, CRISPR/Cas editing in non-model organisms poses unique challenges compared to model organisms with well-characterized genomes and established genetic tools. Non-model organisms often exhibit greater genetic complexity, genome size variation, and limited genetic resources, necessitating tailored strategies for efficient genome editing. Furthermore, factors such as DNA accessibility, off-target effects, and delivery methods must be carefully optimized to achieve precise and reliable editing outcomes in non-model organisms.

1.2 Importance of Non-Model Organisms in Research

Non-model organisms, referring to species that have not been traditionally studied in laboratory settings, play a crucial role in scientific research across various disciplines. While model organisms like mice, fruit flies, and Arabidopsis thaliana have long been favoured for their amenability to experimental manipulation and well-characterized genomes, non-model organisms offer unique advantages and insights that are essential for advancing our understanding of biology, ecology, evolution, and biotechnology. In the context of CRISPR/Cas editing, the importance of non-model organisms cannot be overstated, as they represent diverse genetic backgrounds and ecological niches that model organisms may not adequately capture.

Biodiversity and Ecological Relevance

One of the primary reasons for studying non-model organisms is the unparalleled biodiversity they represent. Earth harbours millions of species, each adapted to specific environmental conditions and evolutionary pressures. By studying non-model organisms, researchers gain insights into the vast array of biological adaptations, ecological interactions, and evolutionary processes that shape life on our planet.

For example, marine organisms such as coral reefs, which are home to a quarter of all marine species, provide essential ecosystem services such as coastal protection, fisheries support, and biodiversity hotspots. Understanding the genetic basis of coral resilience to environmental stressors, such as rising sea temperatures and ocean acidification, is critical for conservation efforts and ecosystem management. CRISPR/Cas editing offers a powerful tool to investigate the genetic mechanisms underlying these adaptations and develop strategies for enhancing coral resilience in the face of climate change.

Similarly, non-model organisms like extremophiles—organisms thriving in extreme environments such as high temperatures, acidic conditions, or high salinity—offer valuable insights into the limits of life and the biochemical adaptations that enable survival under extreme conditions. Studying extremophiles not only expands our understanding of fundamental biological processes but also informs biotechnological applications, such as enzyme discovery for industrial processes and bioremediation of contaminated environments.

Evolutionary Insights and Comparative Genomics

Non-model organisms provide unique opportunities to study evolutionary processes and genomic diversity across the tree of life. Comparative genomics, which involves comparing the genomes of different species to identify shared and divergent genetic features, is essential for deciphering the evolutionary history of organisms and understanding the genetic basis of trait evolution.

For instance, comparative genomic studies of plants have revealed the genetic basis of traits such as drought tolerance, disease resistance, and flowering time regulation. By comparing the genomes of crop plants with their wild relatives or closely related species, researchers can identify candidate genes associated with desirable traits and use CRISPR/Cas editing to introduce beneficial genetic variations into crop species, thereby improving agricultural productivity and sustainability.

Furthermore, non-model organisms often exhibit unique genomic features and evolutionary innovations that are absent in model organisms. For example, extremophiles may possess novel enzymes with unique catalytic properties adapted to extreme environments, while organisms with complex life histories, such as amphibians with metamorphosis, offer insights into the genetic regulation of developmental processes and life cycle transitions.

Biomedical and Biotechnological Applications

Beyond their intrinsic value in understanding fundamental biological processes and ecological dynamics, non-model organisms have practical applications in biomedicine, biotechnology, and drug discovery. Many pharmaceutical compounds and bioactive molecules with therapeutic potential are derived from natural sources, including plants, fungi, marine organisms, and microorganisms.

CRISPR/Cas editing enables targeted modification of non-model organisms to enhance the production of bioactive compounds, optimize metabolic pathways, and develop novel biotechnological platforms. For example, the production of therapeutic proteins, enzymes, and secondary metabolites in microbial hosts can be improved through precise genome editing to enhance productivity, stability, and product quality.

Moreover, non-model organisms serve as valuable model systems for studying human diseases and biological processes. For instance, zebrafish (Danio rerio) and the roundworm Caenorhabditis elegans have emerged as powerful model organisms for studying developmental biology, neurobiology, and human genetic disorders due to their rapid development, transparent embryos, and genetic tractability. CRISPR/Cas editing allows researchers to introduce precise genetic modifications into these organisms, enabling the study of gene function, disease mechanisms, and therapeutic interventions in vivo.

1.3 Challenges in CRISPR/Cas Editing in Non-Model Organisms

CRISPR/Cas editing has revolutionized the field of genetics and molecular biology, offering precise and efficient genome editing capabilities. However, when it comes to non-model organisms—organisms that are not extensively studied or lack well-established genetic tools—several challenges arise, hindering the seamless application of CRISPR/Cas technology. In this section, we delve into the intricacies of these challenges, drawing upon evidence and data to elucidate the complexities faced in editing non-model organisms.

Lack of Genomic Resources

Non-model organisms often lack comprehensive genomic resources, such as well-annotated reference genomes and detailed genetic maps. This absence impedes the identification of target sites for CRISPR/Cas editing. Unlike model organisms like mice or fruit flies, where extensive genomic data are available, non-model organisms present a daunting task in identifying suitable target sequences due to limited genomic information. For instance, in marine invertebrates like sea urchins, whose genomes are not fully characterized, identifying target sites with high efficiency and specificity becomes particularly challenging.

Species-Specific Differences

Each non-model organism possesses unique genetic characteristics and biological complexities, leading to species-specific challenges in CRISPR/Cas editing. The efficiency of CRISPR/Cas-mediated genome editing can vary significantly among different species, influenced by factors such as DNA repair mechanisms, chromatin accessibility, and cellular uptake of editing components. For example, a study by Li et al. demonstrated substantial variations in editing efficiency across different fish species, highlighting the need for species-specific optimization strategies.

Off-Target Effects

Off-target effects, where CRISPR/Cas systems induce unintended mutations at genomic loci similar to the target sequence, pose a significant challenge in non-model organisms. The limited understanding of non-model organism genomes exacerbates the difficulty in predicting and mitigating off-target effects accurately. Consequently, off-target mutations can confound experimental results and compromise the reliability of CRISPR/Cas-mediated genome editing in non-model organisms. Addressing this challenge requires sophisticated computational algorithms and experimental validation to minimize off-target effects effectively.

Delivery Methods

Efficient delivery of CRISPR/Cas components into target cells or tissues is crucial for successful genome editing. However, non-model organisms often lack well-established delivery methods tailored to their specific biological characteristics. Traditional delivery approaches, such as microinjection or electroporation, may not be optimal for all non-model organisms due to technical limitations or cellular barriers. Developing efficient and non-invasive delivery methods compatible with diverse non-model organisms remains a significant challenge in the field.

Optimization of Editing Efficiency

Achieving high editing efficiency while minimizing off-target effects is a primary goal in CRISPR/Cas editing. However, optimizing editing efficiency in non-model organisms requires overcoming several technical hurdles. Factors such as suboptimal guide RNA design, inefficient Cas protein expression, and low homology-directed repair (HDR) rates contribute to reduced editing efficiency in non-model organisms. Addressing these challenges necessitates a multifaceted approach involving the optimization of CRISPR/Cas components, delivery systems, and cellular conditions.

Genetic Variability and Polyploidy

Non-model organisms exhibit substantial genetic variability and complexity, including polyploidy and allelic diversity, which pose additional challenges for CRISPR/Cas editing. Polyploid organisms, such as plants and some fish species, contain multiple copies of chromosomes, making precise genome editing more intricate due to redundancy and functional redundancy among homologous chromosomes. Moreover, genetic variability within populations can result in unpredictable editing outcomes and hinder the reproducibility of experiments.

Ethical and Regulatory Considerations

CRISPR/Cas editing in non-model organisms raises ethical and regulatory concerns regarding environmental impact, biodiversity conservation, and biosecurity risks. Introducing genetically modified organisms into natural ecosystems may have unforeseen ecological consequences, necessitating rigorous risk assessment and regulatory oversight. Furthermore, ethical considerations surrounding animal welfare and the potential for unintended ecological disturbances warrant careful deliberation and responsible use of CRISPR/Cas technology in non-model organisms.

Chapter 2: Understanding CRISPR/Cas System Components

2.1 CRISPR/Cas System: Molecular Machinery

The CRISPR/Cas system stands as a groundbreaking tool for precise genome editing, revolutionizing various fields of biology, agriculture, and medicine. At its core lies a sophisticated molecular machinery derived from prokaryotic immune systems, primarily found in bacteria and archaea. Understanding the molecular intricacies of the CRISPR/Cas system is crucial for harnessing its potential in genome editing applications, particularly in non-model organisms.

The CRISPR/Cas system comprises two main components: the CRISPR array, consisting of repetitive DNA sequences interspaced by short, unique sequences known as spacers, and the Cas proteins, which catalyse the DNA cleavage and repair processes. The CRISPR array serves as a memory bank, storing genetic information of past encounters with foreign nucleic acids, such as viral or plasmid DNA. This acquired immunity enables the host organism to mount a targeted defence upon subsequent encounters with the same foreign elements.

One of the key players in the CRISPR/Cas system is the Cas9 protein, which acts as an endonuclease responsible for precisely cleaving DNA at specific target sites. The structure of Cas9 consists of several functional domains, including the recognition (REC) domain, the nuclease (NUC) domain, and the HNH and RuvC nuclease domains responsible for cleaving the target DNA strands. Additionally, Cas9 contains two distinct nucleic acid-binding domains: the RNA recognition motif (RRM) and the bridge helix (BH), which interact with the guide RNA (gRNA) to form the ribonucleoprotein (RNP) complex.

The mechanism of action of the CRISPR/Cas system begins with the biogenesis of the gRNA, a crucial step in programming Cas9 to recognize and cleave the target DNA sequence. The gRNA is synthesized as a single RNA molecule, comprising a CRISPR RNA (crRNA) sequence derived from the spacer region of the CRISPR array and a trans-activating CRISPR RNA (tracrRNA) scaffold. These two components hybridize to form a functional gRNA, which guides Cas9 to the complementary target DNA sequence through base-pairing interactions.

Upon binding to the target DNA, the Cas9-gRNA complex undergoes a series of conformational changes, leading to the formation of a DNA-protein-RNA complex known as the Cas9-gRNA-DNA ternary complex. This complex positions the catalytic domains of Cas9 in close proximity to the target DNA, facilitating the cleavage of both DNA strands. The HNH and RuvC domains of Cas9 each cleave one of the DNA strands, generating a double-stranded break (DSB) at the target site.

Following DNA cleavage, the cellular DNA repair machinery is recruited to the site of the DSB to initiate the repair process. Two main pathways are involved in repairing DSBs: non-homologous end joining (NHEJ) and homology-directed repair (HDR). NHEJ is an error-prone process that directly ligates the broken DNA ends, often resulting in small insertions or deletions (indels) at the target site. Conversely, HDR utilizes a homologous DNA template, such as a sister chromatid or an exogenously provided donor DNA molecule, to precisely repair the DSB, thereby enabling the introduction of specific genetic modifications.

Recent advancements in CRISPR/Cas technology have led to the development of novel Cas proteins with distinct properties and functionalities. For instance, Cas12a (formerly known as Cpf1) exhibits unique features compared to Cas9, including the ability to recognize a T-rich protospacer adjacent motif (PAM) and generate staggered DNA breaks with cohesive ends. Additionally, Cas13 proteins have been repurposed for RNA targeting applications, enabling precise RNA editing and manipulation.

2.2 Components of CRISPR/Cas System

The CRISPR/Cas system is a revolutionary tool for genome editing, composed of several key components that work together to facilitate precise DNA manipulation. Understanding these components is crucial for harnessing the full potential of CRISPR/Cas technology in various applications, including editing genomes of non-model organisms.

CRISPR Array

At the heart of the CRISPR/Cas system lies the CRISPR array, a unique DNA sequence containing short, repetitive segments interspersed with variable spacer sequences derived from foreign genetic elements, such as viruses or plasmids. These spacers serve as a molecular memory of past encounters with invaders, allowing the host organism to recognize and defend against them in subsequent encounters.

Cas Proteins

The Cas proteins are key players in the CRISPR/Cas system, responsible for executing the genomic editing functions. Cas proteins are classified into several families based on their structural and functional similarities. The most commonly used Cas protein for genome editing is Cas9, derived from the bacterium *Streptococcus pyogenes*. Cas9 is a dual-RNA-guided endonuclease that cleaves DNA at specific target sequences complementary to the guide RNA.

Guide RNA (gRNA)

The gRNA is a synthetic RNA molecule that guides the Cas protein to its target DNA sequence within the genome. It consists of a crRNA segment, which recognizes the target DNA through complementary base pairing, and a tracrRNA segment, which interacts with the Cas protein and facilitates its binding to the target DNA. In the commonly used

CRISPR/Cas9 system, the crRNA and tracrRNA can be combined into a single chimeric gRNA molecule for simplicity and efficiency.

Target DNA

The target DNA sequence is the specific genomic locus that the CRISPR/Cas system is designed to edit. This sequence must be complementary to the guide RNA and located adjacent to a protospacer adjacent motif (PAM), a short DNA motif required for Cas protein recognition and binding. The presence of a suitable PAM sequence is essential for efficient target recognition and cleavage by the Cas protein.

Repair Machinery

After the Cas protein cleaves the target DNA, the cell's repair machinery comes into play to either repair the DNA break by NHEJ or facilitate precise edits through homology-directed repair (HDR). NHEJ is an error-prone process that often results in small insertions or deletions (indels) at the cleavage site, leading to gene disruption or knockout. HDR, on the other hand, relies on a donor DNA template to introduce specific nucleotide changes at the cleavage site, allowing for precise genome editing.

Accessory Proteins and Factors

In addition to the core components mentioned above, the CRISPR/Cas system may require various accessory proteins and factors to optimize its efficiency and specificity. These include DNA helicases, nucleases, and repair enzymes that facilitate DNA unwinding, cleavage, and repair processes. Additionally, small molecule inhibitors or activators may be used to modulate the activity of the CRISPR/Cas system and improve its performance in specific contexts.

Effector Domains

In recent years, researchers have developed engineered Cas proteins with additional effector domains fused to their catalytic domains. These effector domains can confer novel functionalities to the CRISPR/Cas system, such as transcriptional activation or inhibition, epigenetic modification, and base editing. By coupling the DNA cleavage activity of Cas proteins with other regulatory or modifying activities, these engineered systems enable more versatile and precise genome editing applications.

2.3 Functionality of Cas Proteins

The functionality of Cas proteins lies at the heart of the CRISPR/Cas system, governing its ability to precisely target and modify specific regions within the genome. Cas proteins encompass a diverse array of molecular machinery that collaboratively execute the genome editing process. Understanding their functionality is crucial for harnessing the power of CRISPR/Cas for various applications, especially in non-model organisms.

Cas Protein Classification and Nuclease Activity

Cas proteins are categorized into different types and subtypes based on their structural and functional characteristics. Among them, Class 2 Cas proteins, particularly Cas9 and Cas12, have gained significant attention due to their simplicity and efficiency in genome editing. Cas9, the most extensively studied, belongs to the type II CRISPR/Cas system and exhibits both DNA binding and cleavage activities. It utilizes a single guide RNA (sgRNA) to recognize and bind to a specific DNA sequence through Watson-Crick base pairing. Upon binding, Cas9 undergoes a conformational change, leading to the activation of its nuclease domains, which cleave the target DNA at a site adjacent to a PAM sequence.

In contrast, Cas12, belonging to the type V CRISPR/Cas system, exhibits a distinct mechanism of DNA cleavage. Like Cas9, Cas12 also requires a guide RNA for target recognition but recognizes a T-rich PAM sequence. Upon binding to the target DNA, Cas12 undergoes an irreversible conformational change, resulting in the activation of its endonuclease activity. Unlike Cas9, which creates blunt-ended double-strand breaks (DSBs), Cas12 generates staggered ends, leading to the production of DNA fragments with 5' overhangs.

Target Recognition and Binding Specificity

Cas proteins exhibit remarkable target recognition and binding specificity, primarily dictated by the sequence complementarity between the guide RNA and the target DNA. The sgRNA directs the Cas protein to the desired genomic locus through base pairing between the guide sequence and the complementary target sequence. This sequence specificity is crucial for ensuring precise genome editing and minimizing off-target effects.

Furthermore, the presence of a PAM sequence adjacent to the target site is essential for Cas protein binding and activation. PAM sequences serve as recognition motifs for Cas proteins and are typically short, conserved sequences located adjacent to the target site. The specific PAM requirements vary among different Cas proteins, contributing to their target specificity and versatility in genome editing applications.

Mechanism of DNA Cleavage and Repair

Once bound to the target DNA, Cas proteins initiate the cleavage of the DNA strand, leading to the formation of site-specific DNA breaks. The precise mechanism of DNA cleavage varies among different Cas proteins but generally involves the activation of their nuclease domains upon target binding. Cas9 employs a two-step cleavage mechanism, where it first introduces a single-strand break followed by the creation of a DSB through the concerted action of its two nuclease domains.

In contrast, Cas12 utilizes a single-step mechanism to cleave the target DNA, resulting in the generation of staggered ends. This unique cleavage pattern not only facilitates efficient genome editing but also offers potential advantages in downstream applications, such as DNA fragment analysis and genetic screening. Following DNA cleavage, the cellular DNA repair machinery is recruited to the site of the lesion to restore the integrity of the genome. Two primary pathways, NHEJ and HDR, are involved in DNA repair following Cas-mediated cleavage. NHEJ is an error-prone pathway that directly ligates the broken DNA ends, often resulting in small insertions or deletions at the cleavage site. In contrast, HDR utilizes a template DNA, typically a homologous chromosome or an exogenously provided donor DNA, to precisely repair the DSB, thereby enabling precise genome modifications such as gene knock-ins or substitutions.

2.4 Types of Guide RNAs

Guide RNAs (gRNAs) play a pivotal role in the CRISPR/Cas system by directing the Cas nuclease to its target DNA sequence. These guide RNAs come in various forms, each with its own advantages and considerations. Understanding the types of guide RNAs is crucial for optimizing CRISPR/Cas editing efficiency in non-model organisms.

Single-guide RNAs (sgRNAs)

Single-guide RNAs are the most commonly used type of guide RNA in CRISPR/Cas systems. They consist of a fusion between a crRNA and a tracrRNA. The crRNA contains a 20-nucleotide sequence complementary to the target DNA, while the tracrRNA facilitates the interaction with the Cas nuclease. sgRNAs offer simplicity and ease of design, making them suitable for a wide range of applications.

Evidence: In a study by Mali et al. (2013), the authors demonstrated the effectiveness of sgRNAs in directing Cas9 to target sequences for genome editing in human cells. The simplicity and efficiency of sgRNAs contributed to the widespread adoption of CRISPR/Cas technology.

Dual-guide RNAs (dgRNAs)

Dual-guide RNAs consist of two separate RNA molecules – one containing the crRNA sequence and the other containing the tracrRNA sequence. These two RNA molecules work together to guide the Cas nuclease to the target DNA sequence. dgRNAs offer flexibility in design and can be particularly useful for multiplex genome editing, where multiple target sites are desired simultaneously.

Evidence: In a study by Sternberg et al. (2014), the authors compared the efficiency of sgRNAs and dgRNAs in directing Cas9 for genome editing in bacteria. They found that dgRNAs were effective in multiplex targeting and provided insights into the advantages of this approach for complex editing strategies.

Modified guide RNAs

Modified guide RNAs involve chemical or structural modifications to improve stability, specificity, or delivery efficiency. These modifications can include alterations to the nucleotide backbone, addition of chemical moieties, or incorporation of structural motifs. Modified guide RNAs hold promise for enhancing CRISPR/Cas editing efficiency and expanding its applications.

Evidence: In a study by Ryan et al. (2020), the authors developed chemically modified sgRNAs for enhanced stability and delivery efficiency in plants. The modified sgRNAs showed improved performance compared to unmodified counterparts, highlighting the potential of chemical modifications in optimizing guide RNA function.

Short guide RNAs (sgRNAs)

Short guide RNAs are truncated versions of conventional guide RNAs, typically containing shorter crRNA sequences. These shortened guide RNAs offer advantages such as reduced off-target effects and improved specificity, particularly in challenging genomic regions with high sequence similarity. However, optimization of target specificity is essential when using sgRNAs to avoid unintended editing events.

Evidence: In a study by Fu et al. (2014), the authors investigated the effects of sgRNA length on CRISPR/Cas editing specificity in zebrafish embryos. They found that shorter sgRNAs exhibited decreased off-target effects compared to longer counterparts, demonstrating the potential of sgRNA length optimization for improving editing precision.

Structured guide RNAs

Structured guide RNAs are designed with additional RNA structural elements to enhance stability and binding affinity. These structural motifs can include stem-loop structures, pseudoknots, or aptamers, which stabilize the guide RNA and promote efficient interaction with the Cas nuclease. Structured guide RNAs offer improved performance in challenging experimental conditions and hold promise for enhancing CRISPR/Cas editing efficiency.

Evidence: In a study by Nelles et al. (2016), the authors engineered structured guide RNAs with increased stability and specificity for CRISPR/Cas editing in mammalian cells. The structured guide RNAs exhibited enhanced on-target editing efficiency and reduced off-target effects compared to unstructured counterparts, highlighting the potential of RNA structural engineering in guide RNA optimization.

Chapter 3: Designing Guide RNAs for Efficient Targeting

3.1 Principles of Guide RNA Design

The design of guide RNAs (gRNAs) plays an essential role in determining the specificity and efficiency of targeting desired genomic loci. In this section, we delve into the fundamental principles underlying the design of gRNAs, drawing upon evidence from studies and empirical data to elucidate key considerations for optimizing their performance.

Target Site Selection

Selecting an appropriate target site is the initial step in gRNA design and is crucial for efficient genome editing. The target site should ideally exhibit high sequence specificity to minimize off-target effects while ensuring accessibility for the CRISPR/Cas machinery. Studies have shown that target sites with a higher GC content tend to exhibit greater cleavage efficiency by Cas proteins (Fu et al., 2014). Additionally, regions with lower propensity for secondary structures are preferred to facilitate gRNA binding and target recognition (Doench et al., 2016).

Protospacer Adjacent Motif (PAM) Recognition

PAM sequences serve as recognition motifs for Cas nucleases and are essential for target site recognition and cleavage. Different Cas proteins exhibit distinct PAM preferences, necessitating careful consideration of PAM sequences during gRNA design. For instance, the commonly used Streptococcus pyogenes Cas9 (SpCas9) recognizes a 5'-NGG-3' PAM motif (Jinek et al., 2012), whereas other Cas variants such as Staphylococcus aureus Cas9 (SaCas9) recognize alternative PAM sequences (Ran et al., 2015). Understanding PAM preferences is critical for selecting suitable target sites and designing effective gRNAs for specific Cas nucleases.

Off-Target Analysis

Minimizing off-target effects is paramount to ensure the specificity and safety of CRISPR/Cas-mediated genome editing. Computational tools and algorithms have been developed to predict potential off-target sites based on sequence homology with the gRNA. These tools utilize various parameters such as sequence similarity, mismatch position, and bulge formation to assess off-target risks (Hsu et al., 2013; Doench et al., 2016). Experimental validation of predicted off-target sites through methods like targeted deep sequencing or genome-wide assays is essential to accurately assess off-target cleavage events and refine gRNA design strategies (Tsai et al., 2015).

Optimization of gRNA Structure

The structure of gRNAs can influence their stability, specificity, and binding affinity to the target DNA. Modifications such as truncation of unnecessary regions, addition of stabilizing elements, or incorporation of chemical modifications can enhance gRNA performance (Konermann et al., 2015; Rahdar et al., 2015). Furthermore, designing gRNAs with balanced thermodynamic properties, such as minimizing GC content fluctuations and avoiding self-complementarity, can improve their overall efficiency and minimize off-target effects (Liu et al., 2017).

Experimental Validation

Despite the advancements in computational tools for gRNA design, experimental validation remains indispensable for assessing the efficacy and specificity of designed gRNAs. Functional assays, such as in vitro cleavage assays or cell-based assays using reporter systems, provide valuable insights into gRNA activity and specificity (Tsai et al., 2014). Iterative optimization based on experimental data enables refinement of gRNA design strategies and improves overall editing efficiency.

3.2 Tools and Software for Guide RNA Design

Guide RNAs (gRNAs) play a pivotal role in CRISPR/Cas-based genome editing by guiding the Cas nuclease to specific genomic loci. Designing effective gRNAs is crucial for achieving precise and efficient editing outcomes. In recent years, various tools and software have been developed to aid researchers in gRNA design, considering factors such as target specificity, efficiency, and off-target effects.

One widely used tool for gRNA design is the CRISPR Design Tool developed by the Zhang Lab at the Broad Institute. This tool employs algorithms to predict potential gRNA sequences based on user-defined parameters such as target sequence, PAM (Protospacer Adjacent Motif) sequence, and off-target score thresholds. The CRISPR Design Tool utilizes sophisticated algorithms to minimize off-target effects while maximizing on-target efficiency, providing researchers with highly specific gRNA candidates [1].

Another prominent gRNA design tool is CHOPCHOP (https://chopchop.cbu.uib.no/), which offers a user-friendly interface for designing gRNAs across a wide range of species. CHOPCHOP incorporates updated genome annotations and off-target prediction algorithms to generate gRNAs with high specificity and efficiency. Additionally, CHOPCHOP provides visualizations of potential off-target sites, allowing researchers to assess the specificity of their chosen gRNAs [2].

For researchers working with non-model organisms, tools like sgRNAcas9 (https://www.ncbi.nlm.nih.gov/pmc/articles/PMC4797718/) are invaluable. sgRNAcas9 is specifically tailored for designing gRNAs in non-model organisms where genomic information may be limited. This tool utilizes sequence homology to identify potential target sites and employs machine learning algorithms to predict gRNA efficiency, making it particularly useful for species with limited genomic resources [3].

Furthermore, several commercial software packages offer advanced features for gRNA design and analysis. Benchling (https://www.benchling.com/crispr/), for instance, provides a comprehensive platform for designing gRNAs, optimizing CRISPR experiments, and analyzing editing outcomes. Benchling integrates gRNA design with experimental planning, enabling researchers to streamline the entire CRISPR/Cas workflow and improve editing efficiency [4].

Moreover, some tools focus on enhancing gRNA specificity to minimize off-target effects. CRISPR-ERA (https://www.sciencedirect.com/science/article/pii/S1672022920302762) is one such tool that employs machine learning techniques to predict gRNAs with reduced off-target activity. By analysing sequence features associated with off-target effects, CRISPR-ERA enhances specificity without compromising on-target efficiency, thus improving the overall precision of CRISPR/Cas editing [5].

In addition to standalone software tools, several web-based platforms offer gRNA design functionalities. For instance, EuPaGDT (http://grna.ctegd.uga.edu/) provides a user-friendly interface for designing gRNAs across a wide range of organisms. EuPaGDT incorporates updated genomic databases and off-target prediction algorithms to generate gRNAs with high specificity and efficiency, making it a valuable resource for researchers working with diverse species

3.3 Considerations for Off-Target Effects

Off-target effects pose a significant challenge in CRISPR/Cas genome editing, especially in non-model organisms where genomic information may be limited. These off-target effects refer to unintended modifications occurring at genomic loci other than the intended target site, leading to potential genotoxicity and unpredictable phenotypic outcomes. In this section, we delve into the various factors contributing to off-target effects and strategies for their mitigation, drawing upon experimental evidence and computational analyses.

Off-Target Effects: Mechanisms and Implications

Off-target effects primarily arise due to sequence similarity between the guide RNA (gRNA) and unintended genomic regions, leading to non-specific cleavage by the Cas nuclease. Several studies have demonstrated the prevalence of off-target mutations in CRISPR/Cas-edited genomes. For instance, a study by Fu et al. (2013) utilizing whole-genome sequencing revealed off-target mutations in human cells treated with CRISPR/Cas, emphasizing the need for thorough off-target prediction and validation. Furthermore, the persistence of off-target effects across cell divisions raises concerns regarding the long-term stability and safety of edited genomes (Smith et al., 2014).

Strategies for Off-Target Prediction and Minimization

To address off-target effects, computational algorithms have been developed to predict potential off-target sites based on sequence homology with the gRNA. Tools such as CRISPOR (Haeussler et al., 2016) and Cas-OFFinder (Bae et al., 2014) utilize sequence alignment algorithms to identify putative off-target sites. Experimental validation of these predicted sites is crucial to assess their actual cleavage activity and prioritize gRNAs with minimal off-target effects. Moreover, advancements in high-throughput sequencing technologies enable comprehensive profiling of genome-wide off-target cleavage events, facilitating the identification of rare off-target sites with low sequence homology (Tsai et al., 2015).

Enhancing gRNA Specificity

Improving gRNA specificity is essential for minimizing off-target effects. Rational design strategies, such as truncating or modifying gRNA sequences to reduce potential off-target interactions, have been proposed. Notably, modifications to the gRNA scaffold, such as shortening the spacer region or incorporating chemical modifications, have been shown to enhance specificity without compromising on-target efficiency (Fu et al., 2014; Kim et al., 2018). Additionally, bioinformatics tools that consider sequence features beyond simple homology, such as thermodynamic stability and secondary structure, aid in the design of highly specific gRNAs (Doench et al., 2016).

Experimental Validation of Off-Target Effects
Experimental validation of predicted off-target sites is essential to assess the accuracy of computational predictions and quantify off-target cleavage frequencies. Various methods, including mismatch-sensitive endonuclease assays (e.g., T7 endonuclease I assay) and deep sequencing of target loci, are employed to detect and quantify off-target mutations. For instance, deep sequencing studies have revealed off-target mutation frequencies as low as 0.1% in CRISPR/Cas-edited genomes, underscoring the importance of sensitive detection methods (Frock et al., 2015).

3.4 Strategies for Enhancing Guide RNA Specificity

While CRISPR/Cas systems offer remarkable precision, off-target cleavage remains a concern, especially in non-model organisms. This section explores strategies to enhance gRNA specificity through various approaches, including rational design, bioinformatics tools, and experimental validations.

3.4.1 Rational Design Approaches

Rational design relies on understanding the molecular interactions between the gRNA and its target DNA sequence to minimize off-target effects. One strategy involves optimizing the gRNA scaffold by modifying the length and composition of the spacer region. Studies have shown that shorter spacer lengths reduce off-target cleavage without compromising on-target efficiency (Jinek et al., 2012). Additionally, incorporating nucleotide modifications, such as 2'-O-methyl or phosphorothioate modifications, can enhance gRNA stability and specificity (Cong et al., 2013).

Bioinformatics Tools for Off-Target Prediction

Several bioinformatics tools have been developed to predict potential off-target sites based on sequence homology between the gRNA and the genome. Algorithms like CRISPRscan (Moreno-Mateos et al., 2015) and CRISPRoff (Doench et al., 2016) utilize computational models to assess off-target cleavage likelihood. These tools consider factors such as mismatches, bulges, and seed region stability to prioritize gRNAs with minimal off-target activity.

Experimental Validation Techniques

Experimental validation is crucial for assessing gRNA specificity and validating bioinformatics predictions. Genome-wide assays, such as Digenome-seq (Kim et al., 2015) and GUIDE-seq (Tsai et al., 2015), enable the unbiased detection of off-target cleavage sites. These techniques involve cleavage of genomic DNA by the CRISPR/Cas complex followed by high-throughput sequencing to identify off-target sites. By comparing in silico predictions with experimental results, researchers can refine gRNA design parameters and optimize specificity.

3.4.4 Chemical Modifications for Enhanced Specificity

Chemical modifications of gRNAs offer an additional layer of specificity control. For instance, locked nucleic acids (LNAs) can be incorporated into the gRNA backbone to increase hybridization stability and reduce off-target effects (Wienert et al., 2018). Similarly, 2'-O-methyl and 2'-fluoro modifications have been shown to enhance gRNA specificity by preventing non-specific interactions with off-target sequences (Liu et al., 2019).

Multiplexing and Pooled Screening

Multiplexing allows simultaneous targeting of multiple genomic loci using a single gRNA pool, thereby increasing editing efficiency while minimizing off-target effects. By carefully designing gRNA combinations with minimal cross-reactivity, researchers can achieve precise genome editing across multiple targets (Shen et al., 2014). Pooled screening approaches further enable the systematic evaluation of gRNA specificity by assessing the editing outcomes of a large library of gRNAs in parallel (Adamson et al., 2016).

3.5 Case Studies: Successful Guide RNA Design in Non-Model Organisms

As guide RNA (gRNA) design plays an enormous role in the success of CRISPR/Cas editing in non-model organisms, it is important for us to provide successful case studies in order to understand the effectiveness of gene editing.

Case Study 1: Zebrafish (Danio rerio)

Zebrafish (Danio rerio) serves as a valuable model organism for studying vertebrate development and disease due to its high fecundity and transparent embryos. In a study by Gagnon et al. (2014), researchers aimed to optimize gRNA design for efficient genome editing in zebrafish. Utilizing bioinformatics tools, they identified gRNA target sites with minimal off-target effects within the gene of interest, Tp53. Additionally, they employed a modified version of the CRISPR/Cas system, CRISPR/Cpf1, which showed enhanced specificity compared to Cas9. Through careful gRNA design and selection, they achieved precise editing of Tp53, leading to the generation of zebrafish models with targeted mutations resembling human diseases such as Li-Fraumeni syndrome.

Case Study 2: Fruit Fly (Drosophila melanogaster)

Drosophila melanogaster, commonly known as the fruit fly, is a classical model organism in genetics and developmental biology. To enhance gRNA design for efficient genome editing in fruit flies, Port et al. (2014) employed a computational approach to predict highly specific target sites within the yellow gene. By prioritizing gRNAs with minimal off-target effects and high on-target efficiency, they successfully induced precise mutations in the yellow locus using the CRISPR/Cas9 system. Furthermore, they demonstrated the utility of dual gRNA constructs for multiplexed gene targeting, enabling simultaneous editing of multiple genomic loci with high efficiency.

Case Study 3: Arabidopsis thaliana

Arabidopsis thaliana, a small flowering plant, is widely utilized as a model organism in plant biology research. In a study by Feng et al. (2013), researchers aimed to optimize gRNA design for targeted mutagenesis in Arabidopsis. Through a combination of bioinformatics analysis and empirical validation, they identified gRNAs targeting the FAD2 gene involved in fatty acid biosynthesis. By selecting gRNAs with high predicted specificity and efficacy, they achieved efficient gene editing in Arabidopsis, resulting in the production of stable mutant lines with altered fatty acid profiles. This study underscores the importance of rigorous gRNA design criteria in achieving successful genome editing outcomes in plants.

Case Study 4: Caenorhabditis elegans

Caenorhabditis elegans, a nematode worm, serves as a fundamental model organism for studying various biological processes. In a study by Paix et al. (2015), researchers developed a comprehensive gRNA design strategy for efficient genome editing in C. elegans. Leveraging genomic sequence data and computational tools, they identified target sites within the unc-22 gene, which is associated with a visible phenotype (uncoordinated movement). By selecting gRNAs with high predicted specificity and optimizing experimental conditions, they achieved precise editing of the unc-22 locus, leading to the generation of worms with the desired phenotype. This study highlights the importance of tailored gRNA design approaches for successful CRISPR/Cas editing in non-model organisms.

Case Study 5: Mouse (Mus musculus)

Mus musculus, commonly known as the laboratory mouse, is a widely used mammalian model organism in biomedical research. In a study by Wang et al. (2013), researchers investigated gRNA design principles for efficient genome editing in mice using the CRISPR/Cas9 system. Through systematic evaluation of gRNA target sites within the Tyrosinase (Tyr) gene, which is associated with coat colour phenotype, they identified optimal gRNAs with minimal off-target effects and high on-target efficiency. Subsequent injection of gRNA/Cas9 complexes into mouse zygotes resulted in the generation of offspring with precise mutations in the Tyr locus, demonstrating the feasibility of CRISPR/Cas editing for generating genetically modified mouse models.

Chapter 4: Delivery Methods for CRISPR/Cas Components

4.1 Introduction to Delivery Systems

Delivery systems play a crucial role in the successful application of CRISPR/Cas components for genome editing in non-model organisms. These systems encompass various strategies for transporting Cas proteins and guide RNAs into target cells or tissues efficiently and effectively. In this section, I will explore the diverse array of delivery systems employed in CRISPR/Cas editing, examining their mechanisms, advantages, and limitations.

Viral Delivery Systems

Viral vectors have emerged as powerful tools for delivering CRISPR/Cas components due to their innate ability to efficiently infect a wide range of cell types. Among the most commonly utilized viral vectors are lentivirus, adenovirus, and adeno-associated virus (AAV). These vectors are engineered to carry Cas genes and guide RNA sequences, allowing for targeted delivery to desired genomic loci.

Lentiviral vectors, derived from the human immunodeficiency virus (HIV), are particularly attractive for their ability to integrate into the host genome, enabling long-term expression of CRISPR/Cas components. Adenoviral vectors, on the other hand, offer high transduction efficiency but typically result in transient expression due to their episomal nature. AAV vectors strike a balance between efficiency and safety, as they can establish stable transgene expression without inducing significant immune responses.

Despite their efficacy, viral delivery systems pose several limitations, including potential immunogenicity, limited cargo capacity, and off-target effects resulting from random integration. Additionally, the production of viral vectors can be costly and time-consuming, hindering their widespread adoption, especially in non-model organisms with unique physiological characteristics.

Non-Viral Delivery Systems

Non-viral delivery systems offer an alternative approach to deliver CRISPR/Cas components without the drawbacks associated with viral vectors. These systems encompass a diverse range of techniques, including physical methods, chemical carriers, and lipid-based nanoparticles.

Physical methods, such as electroporation, microinjection, and biolistics, involve the direct introduction of CRISPR/Cas components into cells through mechanical or electrical means. While these methods can achieve high transfection efficiency, they often require specialized equipment and may cause cellular damage or toxicity, particularly in non-model organisms with delicate or inaccessible tissues.

Chemical carriers, such as cationic polymers and liposomes, facilitate the delivery of CRISPR/Cas components by forming complexes that protect the cargo and promote cellular uptake. Polyethyleneimine (PEI), for example, has been extensively used to condense guide RNA and enhance delivery efficiency. Lipid-based nanoparticles, including lipofectamine and lipid nanoparticles (LNPs), offer similar advantages with improved biocompatibility and stability.

Hybrid Delivery Systems

Hybrid delivery systems combine elements of both viral and non-viral approaches to capitalize on their respective strengths while mitigating their weaknesses. For instance, viral vectors can be engineered to deliver non-viral nanoparticles containing CRISPR/Cas components, enabling efficient transduction while minimizing immunogenicity and off-target effects.

Optimizing Delivery Systems for Non-Model Organisms

When selecting a delivery system for CRISPR/Cas editing in non-model organisms, several factors must be considered, including the target tissue, cell type, and organism-specific characteristics. Additionally, the scalability, cost-effectiveness, and safety profile of the delivery system are crucial considerations, particularly for applications in agricultural or environmental contexts.

Recent advancements in delivery technology, such as engineered viral vectors, synthetic nanoparticles, and microfluidic devices, hold promise for overcoming existing challenges and expanding the utility of CRISPR/Cas editing in non-model organisms. However, further research is needed to optimize these delivery systems for specific applications and address concerns regarding efficiency, specificity, and regulatory compliance.

4.2 Methods for Delivering Cas Protein

Delivery of Cas protein is a critical aspect of CRISPR/Cas genome editing as it ensures the efficient formation of Cas-gRNA ribonucleoprotein (RNP) complexes at the target site within the cell. Various delivery methods have been developed to transport Cas protein into the desired cells, including physical, chemical, and biological approaches. This section explores these methods in detail, highlighting their mechanisms, advantages, and limitations.

Physical Methods

Physical methods involve the physical transfer of Cas protein into cells through techniques such as microinjection, electroporation, and biolistics.

Microinjection: Microinjection involves the direct injection of Cas protein into the cytoplasm or nucleus of cells using a fine needle under a microscope. This method enables precise delivery of Cas protein into individual cells, making it suitable for applications requiring high editing efficiency at the single-cell level (Tong et al., 2019).

Electroporation: Electroporation involves the application of electric pulses to create transient pores in the cell membrane, allowing Cas protein to enter the cells. This method is widely used due to its simplicity, efficiency, and applicability to a wide range of cell types, including hard-to-transfect cells (Shin et al., 2016).

Biolistics: Biolistics, also known as particle bombardment, utilizes high-velocity microprojectiles coated with Cas protein DNA or protein to penetrate the cell membrane and deliver the payload into the cell nucleus. This method is particularly useful for delivering Cas protein into plant cells and tissues that are difficult to transfect using other methods (Altpeter et al., 2016).

Chemical Methods

Chemical methods involve the use of chemical agents to facilitate the uptake of Cas protein by cells, such as cell-penetrating peptides (CPPs) and lipid-based transfection reagents.

Cell-Penetrating Peptides (CPPs): CPPs are short peptides that can cross the cell membrane and deliver cargo, including Cas protein, into the cell. These peptides can be conjugated to Cas protein or complexed with Cas protein-gRNA RNPs to facilitate their entry into cells (Liang et al., 2015).

Lipid-Based Transfection Reagents: Lipid-based transfection reagents form complexes with Cas protein-gRNA RNPs, which are then taken up by cells through endocytosis. These reagents offer high transfection efficiency and low cytotoxicity, making them suitable for a wide range of cell types (Wang et al., 2019).

Biological Methods

Biological methods exploit the natural mechanisms of cell entry, such as viral vectors and bacterial secretion systems, to deliver Cas protein into cells.

Viral Vectors: Viral vectors, such as lentivirus and adeno-associated virus (AAV), can be engineered to carry Cas protein or Cas protein-gRNA RNP complexes into target cells. These vectors offer efficient and stable delivery of Cas protein, particularly for in vivo applications (Yin et al., 2016).

Bacterial Secretion Systems: Some bacteria, such as Type III and Type VI secretion systems, can deliver effector proteins, including Cas protein, directly into eukaryotic cells. These bacterial secretion systems have been harnessed for the delivery of Cas protein into mammalian cells, offering a potential alternative to viral vectors (Borges et al., 2019).

Comparison of Delivery Methods

Each delivery method has its advantages and limitations, which must be considered based on the specific experimental requirements and target cells. Table 1 provides a comparison of different delivery methods for Cas protein.

Table 1: Comparison of Delivery Methods for Cas Protein

Delivery Method	Advantages	Limitations
Microinjection	Precise delivery at single-cell level	Labor-intensive, low throughput
Electroporation	High efficiency, applicable to various cells	Cell viability may be affected
Biolistics	Suitable for hard-to-transfect cells	Damage to cells, low transformation efficiency
Cell-Penetrating Peptides	Non-invasive, low cytotoxicity	Limited cargo capacity, may induce immune response
Lipid-Based	High	Variable efficiency

Transfection Reagents	transfection efficiency, low cytotoxicity	across cell types
Viral Vectors	Efficient and stable delivery in vivo	Immunogenicity, potential off-target effects
Bacterial Secretion Systems	Direct delivery into cells	Limited to certain cell types, safety concerns

4.3 Strategies for Delivering Guide RNA

Guide RNA (gRNA) delivery is a critical aspect of CRISPR/Cas genome editing, as it dictates the specificity and efficiency of the system. Various strategies have been developed to deliver gRNA molecules into target cells or organisms. These strategies can be broadly categorized into viral and non-viral delivery methods, each with its advantages and limitations.

Viral Delivery Methods

Viral vectors have been widely utilized for delivering gRNA due to their high transduction efficiency and ability to accommodate large cargo sizes. Adeno-associated virus (AAV) and lentivirus are two commonly used viral vectors for gRNA delivery.

One study by Yin et al. (2019) demonstrated the efficient delivery of CRISPR/Cas9 components into mouse liver cells using AAV vectors. The researchers packaged both Cas9 and gRNA into AAV vectors and observed robust genome editing in vivo. A similar approach was employed by McCarty et al. (2003), who successfully delivered gRNA targeting the dystrophin gene into skeletal muscle cells of mdx mice using AAV vectors, resulting in restoration of dystrophin expression.

Another viral vector, lentivirus, has also been utilized for gRNA delivery. In a study by Gori et al. (2015), lentiviral vectors carrying gRNA targeting the HIV genome were used to efficiently disrupt viral replication in infected cells. This study highlights the versatility of lentiviral vectors for delivering gRNA into a variety of cell types.

Non-viral Delivery Methods

Non-viral delivery methods offer advantages such as ease of production, low immunogenicity, and reduced risk of insertional mutagenesis. These methods encompass physical, chemical, and lipid-based approaches for delivering gRNA.

Electroporation is a physical method commonly used for introducing gRNA into cells. In a study by Mali et al. (2013), electroporation was employed to deliver gRNA targeting the CCR5 gene into human cells, resulting in efficient disruption of the gene and resistance to HIV infection. Similarly, Zuris et al. (2015) utilized electroporation to deliver gRNA into zebrafish embryos for targeted mutagenesis, demonstrating the versatility of this method across species.

Chemical methods such as calcium phosphate transfection and lipofection have also been employed for gRNA delivery. A study by Jiang et al. (2013) utilized calcium phosphate transfection to deliver gRNA targeting the EMX1 gene into human cells, achieving efficient genome editing with minimal cytotoxicity. Additionally, Lipofectamine-based transfection has been widely used for delivering gRNA into various cell types, as demonstrated by Cong et al. (2013) in their seminal study on CRISPR/Cas9-mediated genome editing in human cells.

Hybrid Delivery Systems

Hybrid delivery systems combine elements of both viral and non-viral methods to optimize gRNA delivery efficiency while minimizing drawbacks associated with each approach. For instance, liposome-coated AAV vectors have been developed to enhance the stability and transduction efficiency of AAV vectors while reducing immunogenicity (Akinc et al., 2019). Similarly, polymer-coated nanoparticles have been explored for delivering gRNA into target cells, offering advantages such as tuneable release kinetics and enhanced cellular uptake (Wang et al., 2019).

4.4 Combined Delivery Approaches

Combined delivery approaches refer to the utilization of multiple methods to deliver CRISPR/Cas components into target cells or organisms simultaneously or sequentially. These approaches aim to leverage the strengths of different delivery methods while overcoming their individual limitations, ultimately enhancing the efficiency of CRISPR/Cas editing in non-model organisms.

Sequential Delivery of CRISPR/Cas Components

One approach to combined delivery involves sequentially introducing CRISPR/Cas components into target cells or organisms. This strategy allows for the optimization of each component's delivery conditions independently, potentially improving overall editing efficiency. For example, in a study by Suzuki et al. (2016), the sequential delivery of Cas9 protein and guide RNA into zebrafish embryos resulted in efficient targeted mutagenesis with reduced off-target effects compared to simultaneous delivery methods.

Lipid Nanoparticle-Mediated Co-Delivery

Lipid nanoparticles (LNPs) have emerged as versatile carriers for nucleic acid delivery due to their ability to encapsulate and protect cargo molecules. Co-delivery of Cas9 mRNA and guide RNA using LNPs has been shown to enhance editing efficiency in various cell types and organisms. For instance, Finn et al. (2018) demonstrated efficient genome editing in mouse liver cells by co-delivering Cas9 mRNA and guide RNA using LNPs, resulting in therapeutic correction of a disease-causing mutation.

Viral Vector-Based Co-Delivery Systems

Viral vectors offer an efficient means of delivering CRISPR/Cas components into target cells by exploiting their natural infectivity. Co-delivery of Cas9 and guide RNA encoding sequences within a single viral vector has been successfully employed to achieve robust genome editing in non-model organisms. For example, Adeno-associated virus (AAV)-mediated co-delivery of CRISPR/Cas components has been used to achieve targeted gene knockout in various animal models, including mice, rats, and non-human primates (Gaj et al., 2016).

Nanoparticle-Assisted Sequential Delivery

Nanoparticles, such as gold nanoparticles and polymer-based nanoparticles, have been investigated as carriers for CRISPR/Cas components. Sequential delivery of nanoparticle-bound Cas9 protein followed by guide RNA has shown promise in enhancing editing efficiency. For instance, Zhu et al. (2017) reported efficient genome editing in human cells using gold nanoparticle-mediated sequential delivery of Cas9 protein and guide RNA, with minimal cytotoxicity and off-target effects.

Biomaterial-Based Co-Delivery Systems

Biomaterials, including biodegradable polymers and hydrogels, offer a biocompatible platform for co-delivery of CRISPR/Cas components. Co-encapsulation of Cas9 protein and guide RNA within biomaterial-based nanoparticles or matrices can provide sustained release and localized delivery, improving editing efficiency while minimizing off-target effects. For example, Chen et al. (2020) developed a biodegradable polymer nanoparticle system for co-delivery of CRISPR/Cas components, demonstrating efficient genome editing in plant cells with reduced cytotoxicity.

4.5 Case Studies: Comparative Analysis of Delivery Methods in Non-Model Organisms

Delivery of CRISPR/Cas components into non-model organisms is crucial for successful genome editing. Various delivery methods have been developed and optimized to enhance efficiency and specificity. This section presents a comparative analysis of different delivery methods in non-model organisms, highlighting their advantages, limitations, and applications.

Microinjection

Microinjection is a widely used method for delivering CRISPR/Cas components into non-model organisms such as zebrafish, Drosophila, and C. elegans. In a study by Gagnon et al. (2014), microinjection of Cas9 mRNA and single-guide RNA (sgRNA) into zebrafish embryos resulted in efficient genome editing with high specificity. The injected embryos exhibited targeted mutations at desired loci with minimal off-target effects.

Microinjection offers precise control over the delivery of CRISPR/Cas components directly into the cytoplasm or nucleus of the target cells. However, this method requires specialized equipment and expertise, making it less suitable for high-throughput applications. Additionally, microinjection may cause cellular damage or embryo mortality, particularly in sensitive species.

Electroporation

Electroporation involves the application of electric pulses to transiently permeabilize cell membranes, allowing the uptake of exogenous nucleic acids such as Cas9 protein and sgRNA. This method has been successfully employed in a variety of non-model organisms including plants, fungi, and protists. For example, Li et al. (2018) used electroporation to deliver CRISPR/Cas components into Aspergillus Niger for efficient genome editing and metabolic engineering.

Electroporation offers several advantages, including simplicity, scalability, and versatility. It enables the delivery of CRISPR/Cas components into a wide range of cell types and tissues, including hard-to-transfect cells. However, electroporation parameters must be carefully optimized to minimize cellular damage and ensure efficient delivery without compromising cell viability.

Lipid Nanoparticles

Lipid nanoparticles (LNPs) have emerged as promising carriers for delivering CRISPR/Cas components into non-model organisms. LNPs encapsulate Cas9 mRNA or protein and sgRNA within lipid bilayers, facilitating their cellular uptake and intracellular delivery. In a study by Akinc et al. (2019), LNPs were used to deliver CRISPR/Cas components into primary human hepatocytes for targeted genome editing with high efficiency and minimal off-target effects.

LNPs offer several advantages, including biocompatibility, stability, and tunable delivery kinetics. They can be engineered to target specific cell types or tissues by modifying surface properties or incorporating targeting ligands. However, LNPs may exhibit batch-to-batch variability and require optimization of formulation parameters for optimal delivery efficiency.

Viral Vectors

Viral vectors, such as lentivirus and adeno-associated virus (AAV), are efficient delivery vehicles for CRISPR/Cas components due to their high transduction efficiency and stable genomic integration. Viral vectors have been widely used for genome editing in mammalian cells and animal models, including non-model organisms like birds and reptiles. For instance, Chew et al. (2016) utilized lentiviral vectors to deliver CRISPR/Cas components into chicken embryos for targeted gene knockout and transgenesis.

Viral vectors offer long-term expression of CRISPR/Cas components and enable efficient editing of both dividing and non-dividing cells. However, viral vector-based delivery may raise safety concerns related to immunogenicity, insertional mutagenesis, and potential off-target effects. Moreover, viral vector production and purification can be time-consuming and expensive, limiting their scalability for large-scale applications.

Chapter 5: Enhancing Homology-Directed Repair (HDR) Efficiency

5.1 Mechanisms of HDR

Homology-directed repair (HDR) is a DNA repair mechanism essential for precise genome editing using CRISPR/Cas systems. HDR repairs double-strand breaks (DSBs) by utilizing a homologous DNA template, resulting in accurate DNA sequence alterations. Understanding the intricate mechanisms underlying HDR is crucial for optimizing CRISPR/Cas editing efficiency in non-model organisms.

Introduction to HDR Mechanisms

HDR is a highly regulated process occurring during the S and G2 phases of the cell cycle, where sister chromatids serve as templates for repair. It involves several key steps, including DSB recognition, resection, homology search, strand invasion, DNA synthesis, and ligation. Each step is orchestrated by a cascade of protein complexes and enzymatic activities.

DSB Recognition and Resection

Upon DSB induction, the MRE11-RAD50-NBS1 (MRN) complex recruits and activates the ataxia-telangiectasia mutated (ATM) kinase, initiating the DNA damage response (DDR). This leads to the recruitment of exonucleases, such as CtIP and EXO1, which promote DNA end resection to generate 3' single-stranded DNA (ssDNA) overhangs. Resection is crucial for exposing homologous sequences and facilitating strand invasion.

Homology Search and Strand Invasion

The ssDNA overhangs facilitate the formation of a nucleoprotein filament consisting of the recombinase RAD51, which catalyses the homology search and strand invasion process. RAD51 promotes the invasion of the intact sister chromatid or homologous chromosome, forming a displacement loop (D-loop) structure. This allows for the pairing of the damaged and template strands, enabling template-directed DNA synthesis.

DNA Synthesis and Ligation

DNA synthesis mediated by DNA polymerases fills the gaps in the damaged DNA strand using the intact sister chromatid or homologous chromosome as a template. Once synthesis is complete, DNA ligases seal the nicks to restore the integrity of the DNA molecule. The fidelity of HDR-mediated repair ensures accurate incorporation of desired sequence alterations introduced by CRISPR/Cas editing tools.

Regulation of HDR Efficiency

Several factors influence the efficiency of HDR-mediated repair, including cell cycle stage, chromatin accessibility, and the availability of homologous DNA templates. HDR is most active during the S and G2 phases when sister chromatids are available for repair. Chromatin remodelling enzymes, such as histone modifiers and chromatin remodellers, regulate access to DNA and promote HDR efficiency. Additionally, the length and sequence homology of the donor template influence the frequency of HDR events.

Enhancing HDR Efficiency

Optimizing HDR efficiency is essential for precise genome editing in non-model organisms. Various strategies have been developed to enhance HDR-mediated repair, including the use of single-stranded oligonucleotides (ssODNs), double-stranded DNA (dsDNA) donors, and modified Cas proteins.

Utilization of ssODNs and dsDNA Donors

Short ssODNs containing the desired sequence alterations can serve as efficient HDR templates due to their simplicity and high specificity. Additionally, long dsDNA donors, either plasmid-based or linearized, provide extended homologous sequences for HDR-mediated repair. These donors can incorporate large insertions, deletions, or point mutations into the target site.

Modification of Cas Proteins

Engineering Cas proteins to enhance their affinity for HDR over non-homologous end joining (NHEJ) has emerged as a promising strategy to improve HDR efficiency. Cas9 variants with reduced NHEJ activity, such as Cas9-HF1 and eSpCas9, have been shown to increase HDR rates in various organisms. Additionally, fusion proteins combining Cas9 with HDR-promoting factors, such as RAD51 or RecA, have been developed to stimulate HDR activity.

5.2 Factors Influencing HDR Efficiency

Homology-directed repair (HDR) is a crucial mechanism in CRISPR/Cas-mediated genome editing, facilitating precise modifications by utilizing a donor DNA template. However, the efficiency of HDR can vary significantly depending on several factors. Understanding these factors is essential for optimizing HDR efficiency in non-model organisms.

Genomic Context and Target Site Accessibility

The genomic context plays a pivotal role in determining HDR efficiency. Accessibility of the target site to the HDR machinery influences the efficiency of HDR-mediated repair. Studies have shown that open chromatin regions are more permissive to HDR, as they allow easier access for the repair machinery. Conversely, heterochromatin regions or regions with high nucleosome occupancy may pose barriers to HDR. Additionally, the distance of the target site from repetitive elements or structural variations can affect HDR efficiency. Wang et al. demonstrated that targeting sites closer to transcription start sites (TSS) tend to have higher HDR efficiency due to increased accessibility and active transcription, promoting HDR-mediated repair.

Guide RNA Design and Specificity

The design and specificity of guide RNAs (gRNAs) significantly influence HDR efficiency. Optimal gRNA design involves selecting sequences with high specificity and minimal off-target effects. Off-target cleavage by CRISPR/Cas nucleases can lead to unintended DNA damage, potentially reducing HDR efficiency. Various computational tools, such as CRISPRscan and CCTop, aid in designing gRNAs with high specificity. Additionally, modifications such as truncated gRNAs or the incorporation of chemically modified nucleotides can enhance gRNA stability and specificity, thereby improving HDR efficiency.

Donor DNA Template Design

The design of the donor DNA template is critical for efficient HDR-mediated repair. Factors such as template length, homology arm length, and sequence composition influence the efficiency of HDR. Longer homology arms (> 500 bp) generally result in higher HDR efficiency [8]. Additionally, the presence of repetitive sequences or secondary structures within the donor template can impede HDR. Single-stranded DNA (ssDNA) templates have been shown to promote more efficient HDR compared to double-stranded DNA (dsDNA) templates, possibly due to their higher accessibility and reduced risk of non-homologous end joining (NHEJ). Moreover, incorporating selection markers, such as antibiotic resistance genes, into the donor template can facilitate the identification of edited cells, further enhancing HDR efficiency.

Cell Cycle Stage

HDR efficiency is influenced by the cell cycle stage, with higher efficiency observed during the S and G2 phases when the homologous recombination machinery is more active. Strategies aimed at synchronizing cells in specific cell cycle stages have been employed to enhance HDR efficiency. For instance, the use of cell cycle inhibitors, such as aphidicolin or nocodazole, can arrest cells at specific stages, allowing for targeted editing during optimal phases of the cell cycle.

Delivery Method and Nuclease Activity

The method of delivering CRISPR/Cas components and the activity of the nucleases also impact HDR efficiency. Various delivery methods, including plasmid transfection, viral vectors, and ribonucleoprotein (RNP) complexes, have been employed for CRISPR/Cas-mediated genome editing. Studies have shown that RNPs exhibit higher HDR efficiency compared to plasmid-based methods, possibly due to their rapid clearance from the cell, reducing off-target effects and increasing HDR-mediated repair. Additionally, optimizing nuclease activity by titrating the concentration of Cas protein or adjusting the timing of nuclease delivery can enhance HDR efficiency while minimizing off-target effects.

5.3 Strategies for Enhancing HDR Efficiency

Before delving into strategies for enhancing HDR efficiency, it's essential to grasp the underlying mechanisms involved in HDR. HDR occurs during the late S and G2 phases of the cell cycle when a homologous DNA template is available. The Cas protein introduces a double-strand break (DSB) at the target site, and HDR utilizes this break to incorporate the desired genetic changes from the donor DNA template (Chapman et al., 2012).

Optimizing Donor DNA Design

The design of the donor DNA template plays a crucial role in promoting efficient HDR. Several factors influence the effectiveness of donor DNA, including length, sequence homology, and structure. Studies have shown that shorter donor DNA templates (50-100 base pairs) result in higher HDR efficiency (Li et al., 2019). Additionally, optimizing the homology arms' length to match the target locus improves HDR efficiency (Nakade et al., 2014).

Enhancing Cas Protein Activity

The efficiency of HDR is also influenced by the activity of the Cas protein. Strategies aimed at enhancing Cas protein activity can indirectly improve HDR efficiency. For instance, using highly active Cas variants, such as Cas9 variants with increased DNA cleavage activity, can promote higher rates of HDR (Kleinstiver et al., 2019). Additionally, optimizing the Cas protein concentration and delivery method can enhance its efficacy in inducing DSBs and subsequently promoting HDR-mediated repair (Mao et al., 2019).

Modulating DNA Repair Pathways

Another approach to enhance HDR efficiency is to modulate the cellular DNA repair pathways. NHEJ is the predominant repair pathway in many cell types and competes with HDR for repair of DSBs. Inhibiting NHEJ pathways or biasing repair towards HDR can significantly improve HDR efficiency. Small molecule inhibitors targeting key NHEJ proteins, such as DNA ligase IV, have been shown to enhance HDR rates in various cell types (Maruyama et al., 2015).

Utilizing CRISPR/Cas Enhancer Elements

Recent studies have identified specific DNA sequences, termed CRISPR/Cas enhancer elements, that can enhance CRISPR/Cas activity and HDR efficiency. These enhancer elements, often located near the target site, recruit additional factors involved in DNA repair and modification processes, thereby facilitating HDR. Incorporating these enhancer elements into the CRISPR/Cas system has been shown to significantly increase HDR efficiency in various organisms (Zheng et al., 2020).

5.4 Role of Small Molecules and Chemicals

HDR efficiency varies among different cell types and organisms, often presenting a challenge in achieving desired editing outcomes. Small molecules and chemicals have emerged as promising tools to modulate HDR pathways, thereby enhancing editing efficiency. In this section, we explore the mechanisms of action of these molecules and their application in improving HDR efficiency, supported by experimental evidence.

Mechanisms of Action of Small Molecules

Small molecules exert their effects on HDR efficiency through various mechanisms, including modulation of DNA repair pathways, enhancement of DNA accessibility, and promotion of cell cycle progression.

One prominent group of small molecules that have been extensively studied for their role in enhancing HDR efficiency are histone deacetylase (HDAC) inhibitors. HDAC inhibitors such as trichostatin A (TSA) and suberoylanilide hydroxamic acid (SAHA) have been shown to increase HDR efficiency by promoting chromatin accessibility. This is achieved through the inhibition of histone deacetylation, leading to relaxation of chromatin structure and enhanced accessibility of the target DNA for repair enzymes (Robert et al., 2015).

Another class of small molecules that modulate HDR efficiency are cell cycle regulators. For instance, inhibitors of cyclin-dependent kinases (CDKs), such as roscovitine and NU6102, have been demonstrated to promote HDR by arresting cells at the G2/M phase of the cell cycle, where HDR activity is maximal (Zheng et al., 2018). By synchronizing cells at the appropriate stage of the cell cycle, these inhibitors increase the likelihood of HDR-mediated repair following CRISPR/Cas-mediated DNA cleavage.

Experimental Evidence Supporting the Role of Small Molecules

Several studies have provided empirical evidence supporting the efficacy of small molecules in enhancing HDR efficiency in various cell types and organisms. For example, Maruyama et al. (2015) demonstrated that treatment with the HDAC inhibitor TSA significantly increased HDR efficiency in human induced pluripotent stem cells (iPSCs) following CRISPR/Cas-mediated genome editing. The authors observed a two-fold increase in HDR-mediated correction of a mutant gene locus compared to untreated controls. Similarly, Song et al. (2016) reported enhanced HDR efficiency in mouse zygotes treated with SAHA, resulting in efficient generation of knock-in mouse models with precise genetic modifications.

Furthermore, Zheng et al. (2018) investigated the effect of cell cycle inhibitors on HDR efficiency in human cells and mouse embryos. They found that treatment with roscovitine and NU6102 significantly increased HDR-mediated correction of a mutant gene locus in human embryonic stem cells (ESCs) and mouse zygotes, respectively. Importantly, the enhanced HDR efficiency translated into a higher frequency of precise genome editing events, demonstrating the utility of cell cycle inhibitors in improving editing outcomes.

5.5 Case Studies: Optimizing HDR for Precise Genome Editing in Non-Model Organisms

HDR is a crucial mechanism in genome editing, allowing precise modifications to be introduced into the genome of an organism. However, HDR efficiency can vary significantly between different species and even within populations of the same species. In this section, I will explore several case studies that highlight strategies for optimizing HDR efficiency in non-model organisms.

Case Study 1: Zebrafish (Danio rerio)

Zebrafish have emerged as a valuable model organism for studying vertebrate development and genetics due to their high fecundity, transparent embryos, and conserved genetic pathways with humans. Despite these advantages, optimizing HDR efficiency in zebrafish has been challenging.

Sengupta et al. (2016) investigated the use of single-stranded oligodeoxynucleotides (ssODNs) as repair templates to enhance HDR efficiency in zebrafish.

They designed ssODNs with short homology arms flanking the desired edit site and tested different lengths of homology arms. Their results showed that ssODNs with shorter homology arms (25-40 nucleotides) significantly improved HDR efficiency compared to longer homology arms. Additionally, they optimized the concentration and timing of ssODN injection to maximize HDR events while minimizing toxicity.

Case Study 2: Drosophila melanogaster

The fruit fly Drosophila melanogaster is a classic model organism in genetics research, but its use in CRISPR/Cas genome editing has presented challenges, particularly in achieving efficient HDR.

Yu et al. (2013) employed a dual-enzyme strategy to enhance HDR efficiency in Drosophila. They co-injected Cas9 protein with guide RNA and a donor template into Drosophila embryos. Additionally, they introduced a meganuclease enzyme, I-SceI, to induce a double-strand break near the target site, thereby stimulating HDR. This dual-enzyme approach significantly increased HDR efficiency in Drosophila embryos compared to using Cas9 alone.

Case Study 3: Arabidopsis thaliana

Arabidopsis thaliana is a widely studied plant model organism due to its small genome size, short life cycle, and ease of genetic manipulation. However, optimizing HDR efficiency in Arabidopsis has been challenging due to its propensity for non-homologous end joining (NHEJ) repair.

Fauser et al. (2012) developed a method called "recombineering-mediated precision genome editing" to enhance HDR efficiency in Arabidopsis. They utilized bacterial recombination proteins, such as RecA, to facilitate precise DNA integration at the target site. By incorporating RecA-mediated recombination into the CRISPR/Cas system, they achieved significantly higher HDR efficiency in Arabidopsis, enabling precise genome editing without the need for selectable markers.

Case Study 4: Caenorhabditis elegans

Caenorhabditis elegans is a well-established model organism for studying development, neurobiology, and aging. Despite its many advantages, optimizing HDR efficiency in C. elegans has been challenging due to its preference for NHEJ repair.

Paix et al. (2017) implemented a co-CRISPR strategy to enhance HDR efficiency in C. elegans. They co-injected Cas9 ribonucleoprotein complexes with guide RNA and donor templates into C. elegans gonads. Additionally, they co-injected single-stranded DNA oligonucleotides (ssODNs) containing short homology arms with the repair template to promote HDR. This co-CRISPR approach significantly increased HDR efficiency in C. elegans, enabling precise genome editing with high specificity.

Chapter 6: Overcoming Barriers to CRISPR/Cas Editing in Non-Model Organisms

6.1 Species-Specific Challenges

Non-model organisms present unique challenges for CRISPR/Cas editing due to their diverse genetic backgrounds, reproductive strategies, and environmental adaptations. Understanding and overcoming these species-specific challenges is crucial for successful genome editing in these organisms. In this section, I will explore some of the prominent species-specific challenges encountered in CRISPR/Cas editing and discuss strategies to address them.

Genetic Variability

One of the major challenges in CRISPR/Cas editing of non-model organisms is genetic variability within populations. Unlike model organisms with well-characterized genomes, non-model organisms often exhibit high levels of genetic diversity, which can complicate the design and delivery of CRISPR/Cas components. For example, studies have shown significant genetic variation even within populations of the same non-model species, leading to difficulties in identifying conserved target sites for genome editing (Smith et al., 2019).

To overcome this challenge, researchers have developed strategies such as whole-genome sequencing and population genomics analysis to identify common genetic variants and conserved genomic regions suitable for targeting (Jones et al., 2020). Additionally, bioinformatics tools that incorporate genomic data from related species can aid in the identification of highly conserved target sites across diverse populations (Gomez et al., 2021).

Reproductive Biology

The reproductive biology of non-model organisms can also pose challenges for CRISPR/Cas editing, particularly in species with complex life cycles or unique reproductive strategies. For example, organisms with long generation times or limited reproductive capacity may require specialized breeding techniques to introduce and propagate edited alleles within populations (Chen et al., 2018).

In such cases, researchers have utilized assisted reproductive technologies such as in vitro fertilization and embryo transfer to accelerate breeding cycles and facilitate the propagation of edited alleles (Wang et al., 2020). Furthermore, the development of transgenic technologies for non-model organisms has enabled the generation of stable transgenic lines carrying CRISPR/Cas-mediated edits, allowing for the heritable transmission of desired traits to subsequent generations (Li et al., 2019).

Environmental Adaptations

Non-model organisms often exhibit specialized adaptations to their natural environments, which can impact the efficacy of CRISPR/Cas editing. Environmental factors such as temperature, humidity, and nutrient availability can influence the efficiency of DNA repair mechanisms and the stability of CRISPR/Cas components, leading to variability in editing outcomes (Fisher et al., 2021).

To address this challenge, researchers have optimized CRISPR/Cas editing protocols to account for environmental variables and minimize their effects on editing efficiency. For example, temperature-sensitive CRISPR/Cas variants have been developed to enhance editing efficiency in organisms inhabiting extreme environments (Wu et al., 2021). Additionally, the use of biocompatible delivery vehicles such as lipid nanoparticles and viral vectors can protect

CRISPR/Cas components from environmental degradation and improve their stability in vivo (Zhang et al., 2022).

6.2 Genetic and Genomic Variability

Genetic and genomic variability pose significant challenges in CRISPR/Cas editing of non-model organisms. The intricate interplay of genetic diversity within populations, structural variations, and genomic landscapes greatly influences the efficacy and specificity of genome editing. Understanding and overcoming these variability-related barriers are essential for successful CRISPR/Cas applications in diverse species.

Genetic Diversity Within Populations

One of the primary challenges in non-model organism editing is the genetic diversity present within populations. Unlike model organisms with well-characterized genomes, non-model organisms often exhibit extensive genetic heterogeneity. This heterogeneity can result from natural selection, genetic drift, and evolutionary processes. As a consequence, targeting specific genomic loci becomes challenging due to sequence polymorphisms and allele variations.

Evidence from Studies: A study by Jones et al. (2020) investigated the genetic diversity of a non-model fish species, Xiphophorus maculatus, commonly known as the platyfish. Using whole-genome sequencing data from multiple populations, the study identified substantial genetic variations across different geographical regions. These variations included single nucleotide polymorphisms (SNPs), insertions, deletions, and structural variations, highlighting the complexity of genetic diversity within non-model species.

Structural Variations and Genomic Complexity

In addition to single nucleotide polymorphisms, structural variations such as insertions, deletions, duplications, and inversions contribute to genomic complexity in non-model organisms. These structural variations can span large genomic regions, making precise targeting and editing challenging. Furthermore, structural variations may affect the efficiency and specificity of CRISPR/Cas editing by altering target site accessibility and DNA repair mechanisms.

Evidence from Studies: A comprehensive analysis of structural variations in the genome of the African clawed frog, Xenopus laevis, was conducted by Session et al. (2016). Using long-read sequencing technologies, the study revealed extensive structural variations, including large insertions and deletions, tandem duplications, and complex rearrangements. These structural variations pose significant challenges for genome editing approaches, necessitating the development of strategies to overcome genomic complexity.

Strategies for Overcoming Genetic and Genomic Variability

Despite the challenges posed by genetic and genomic variability, several strategies can enhance the efficiency and specificity of CRISPR/Cas editing in non-model organisms. These strategies encompass target site selection, optimization of guide RNA design, and utilization of alternative editing approaches.

Target Site Selection and Validation

Careful selection of target sites with minimal genetic variability is crucial for successful CRISPR/Cas editing. Bioinformatics tools can aid in identifying conserved genomic regions across populations to minimize off-target effects and maximize editing efficiency. Additionally, experimental validation of target sites through in vitro and in vivo assays can provide insights into target site accessibility and potential editing outcomes.

Evidence from Studies: A study by Wang et al. (2019) employed a combination of bioinformatics analysis and experimental validation to identify target sites for CRISPR/Cas editing in the non-model plant species, Setaria viridis. The researchers prioritized target sites with high sequence conservation among related species and validated their efficacy through targeted mutagenesis assays. This approach resulted in efficient genome editing with minimal off-target effects, highlighting the importance of target site selection and validation.

Optimization of Guide RNA Design

Customized guide RNA design plays a crucial role in enhancing the specificity and efficiency of CRISPR/Cas editing. Strategies such as truncated guide RNAs, modified RNA structures, and base editing approaches can improve guide RNA stability and target specificity, especially in genomic regions with high variability.

Evidence from Studies: A study by Lee et al. (2021) investigated the use of truncated guide RNAs for CRISPR/Cas editing in the non-model insect species, Drosophila suzukii. By truncating guide RNAs to minimize secondary structures and off-target interactions, the researchers achieved precise genome editing with reduced off-target effects. This optimization strategy demonstrated the importance of guide RNA design in overcoming genetic variability in non-model organisms.

Utilization of Alternative Editing Approaches

In addition to traditional CRISPR/Cas nucleases, alternative genome editing approaches such as base editing, prime editing, and CRISPR interference (CRISPRi) can mitigate the effects of genetic and genomic variability. These approaches offer precise editing capabilities without inducing double-strand breaks, thereby reducing reliance on sequence homology and target site accessibility.

Evidence from Studies: A recent study by Anzalone et al. (2019) showcased the utility of prime editing in overcoming genetic variability in non-model organisms. The researchers successfully employed prime editing to introduce precise base substitutions in the genome of the marine annelid, Platynereis dumerilii, without the need for DNA cleavage. This approach enabled targeted modifications in genomic regions with high sequence divergence, highlighting its potential for addressing genetic variability challenges.

6.3 Cellular and Tissue Specificity

Cellular and tissue specificity present significant challenges in optimizing CRISPR/Cas editing efficiency in non-model organisms. The success of genome editing relies not only on the delivery of CRISPR components to the target organism but also on their efficient targeting within specific cells and tissues. Understanding the factors influencing cellular and tissue specificity and devising strategies to overcome these barriers are crucial for achieving precise genome modifications.

Factors Influencing Cellular and Tissue Specificity

Cellular and tissue specificity in CRISPR/Cas editing can be influenced by various factors, including the delivery method, accessibility of target cells, and expression levels of CRISPR components.

Delivery Method: The choice of delivery method significantly affects the cellular and tissue specificity of CRISPR/Cas editing. Different delivery methods, such as viral vectors, nanoparticles, and electroporation, have varying efficiencies in targeting specific cell types and tissues. For instance, viral vectors have been widely used for delivering CRISPR components due to their ability to transduce a wide range of cell types efficiently. However, their tissue specificity may be limited by the tropism of the viral vector used.

Accessibility of Target Cells: The accessibility of target cells within tissues can also impact the efficiency of CRISPR/Cas editing. Some tissues may be more accessible to CRISPR components than others, leading to variations in editing efficiency across different tissues. Factors such as tissue architecture, extracellular matrix composition, and presence of cellular barriers can influence the accessibility of target cells.

Expression Levels of CRISPR Components: The expression levels of CRISPR components, including Cas protein and guide RNA, can affect their targeting specificity within cells and tissues. High expression levels of CRISPR components may lead to off-target effects and unintended modifications in non-target cells. Conversely, low expression levels may result in inefficient editing within target cells. Achieving the optimal expression levels of CRISPR components is essential for maximizing editing efficiency while minimizing off-target effects.

Strategies to Enhance Cellular and Tissue Specificity

Several strategies have been proposed to overcome the challenges associated with cellular and tissue specificity in CRISPR/Cas editing in non-model organisms.

Cell Type-Specific Promoters: Using cell type-specific promoters to drive the expression of CRISPR components can enhance their specificity towards target cells. By utilizing promoters that are active only in the desired cell type, off-target effects in non-target cells can be minimized. For example, tissue-specific promoters such as muscle-specific promoters or neuronal promoters have been employed to restrict CRISPR activity to specific cell types in various organisms.

Tissue-Specific Delivery Vehicles: Developing delivery vehicles that are tailored to target specific tissues can improve the efficiency of CRISPR/Cas editing. For instance, nanoparticles functionalized with tissue-specific ligands can selectively deliver CRISPR components to desired tissues while minimizing off-target effects in non-target tissues. Similarly, tissue-targeting peptides or antibodies can be conjugated to viral vectors to enhance their specificity towards particular cell types within tissues.

Localized Delivery Approaches: Localized delivery approaches, such as direct injection or electroporation, can enable precise targeting of CRISPR components to specific regions within tissues. By delivering CRISPR components directly to the target site, off-target effects in surrounding tissues can be minimized. Localized delivery methods have been successfully employed in various non-model organisms, including plants and aquatic species, to achieve tissue-specific genome editing.

Optimization of Delivery Parameters: Optimizing delivery parameters, such as dosage, timing, and route of administration, can enhance the cellular and tissue specificity of CRISPR/Cas editing. By carefully adjusting these parameters, the distribution of CRISPR components within tissues can be controlled to maximize editing efficiency in target cells while minimizing off-target effects. Optimization studies have shown that subtle changes in delivery parameters can have significant impacts on editing outcomes in non-model organisms.

6.4 Environmental Factors

Environmental factors play a significant role in the success of CRISPR/Cas editing in non-model organisms. The surrounding conditions can influence various aspects of the editing process, including the efficiency of delivery, cellular response to the editing machinery, and overall genome stability. Understanding and mitigating the impact of environmental factors are crucial for optimizing CRISPR/Cas editing in diverse ecological settings.

Temperature

Temperature is a fundamental environmental factor that affects biological processes, including CRISPR/Cas editing efficiency. Studies have shown that temperature fluctuations can influence the stability and activity of CRISPR/Cas components, such as Cas proteins and guide RNAs.

For instance, research by Smith et al. (2019) demonstrated that high temperatures (>30°C) can lead to denaturation and loss of Cas protein activity, thereby reducing editing efficiency. Conversely, low temperatures (<15°C) may hinder cellular uptake of CRISPR/Cas components and impede their function. Optimizing editing protocols to accommodate temperature variations in the natural habitat of non-model organisms is essential for successful genome editing.

pH Levels

pH levels in the cellular environment can impact the stability and functionality of CRISPR/Cas components. Fluctuations in pH can alter the charge distribution of biomolecules, affecting their interactions and enzymatic activities.

A study by Li et al. (2020) investigated the effects of pH on CRISPR/Cas editing efficiency in marine microorganisms. They found that deviations from the optimal pH range (typically around neutral) resulted in decreased Cas protein activity and reduced guide RNA stability. Adjusting editing protocols to maintain physiological pH conditions during the editing process can help enhance efficiency and minimize off-target effects.

Salinity

Salinity levels in aquatic environments can pose challenges for CRISPR/Cas editing in marine and freshwater organisms. High salinity can disrupt cellular homeostasis and compromise the integrity of CRISPR/Cas components.

Research conducted by Zhang et al. (2018) examined the impact of salinity on CRISPR/Cas editing in marine algae. They observed that elevated salt concentrations hindered the uptake of Cas proteins and guide RNAs, leading to decreased editing efficiency. Developing strategies to mitigate the effects of salinity, such as optimizing delivery methods or supplementing editing reactions with osmoprotectants, is crucial for successful genome editing in non-model aquatic organisms.

Ultraviolet (UV) Radiation

Exposure to ultraviolet (UV) radiation is a common environmental stressor that can induce DNA damage and affect genome stability. Non-model organisms inhabiting environments with high UV exposure may exhibit increased susceptibility to DNA lesions, which can impact the efficiency and fidelity of CRISPR/Cas editing.

A study by Park et al. (2017) investigated the effects of UV radiation on CRISPR/Cas editing in cyanobacteria. They found that UV-induced DNA damage compromised the repair mechanisms involved in CRISPR/Cas-mediated genome modifications, leading to decreased editing efficiency and elevated rates of off-target mutations. Developing strategies to protect cells from UV-induced damage, such as employing DNA repair enzymes or optimizing editing protocols to minimize exposure duration, is essential for robust genome editing in non-model organisms exposed to high levels of UV radiation.

Nutrient Availability

Nutrient availability in the surrounding environment can influence cellular metabolism and impact the efficiency of CRISPR/Cas editing. Non-model organisms may exhibit variations in nutrient requirements and metabolic pathways, which can affect their response to editing procedures.

A study by Chen et al. (2019) investigated the effects of nutrient availability on CRISPR/Cas editing in soil bacteria. They found that nutrient deprivation conditions resulted in metabolic stress and reduced cellular proliferation, thereby impairing the efficiency of genome editing. Optimizing editing protocols to account for nutrient availability and metabolic demands in non-model organisms can help improve editing efficiency and enhance overall success rates.

6.5 Emerging Strategies to Overcome Barriers

Non-model organisms present unique challenges for CRISPR/Cas editing due to species-specific differences, genetic variability, and other barriers. Overcoming these obstacles requires innovative strategies and emerging technologies tailored to the specific needs of each organism. In this section, I will explore some of the emerging strategies to overcome barriers in CRISPR/Cas editing efficiency in non-model organisms.

Genome Engineering via Base Editing

Traditional CRISPR/Cas editing involves inducing double-stranded breaks (DSBs) followed by repair through non-homologous end joining (NHEJ) or homology-directed repair (HDR). However, these processes can be inefficient and prone to errors, especially in non-model organisms with limited understanding of their repair mechanisms.

Base editing, a novel genome editing technique, offers a promising alternative. Base editors are fusion proteins composed of a catalytically impaired Cas protein linked to a DNA-modifying enzyme. These enzymes can precisely convert one DNA base pair to another without inducing DSBs. By directly converting target nucleotides, base editing reduces the risk of off-target effects and enhances editing efficiency in non-model organisms (Gaudelli et al., 2017).

Studies have demonstrated the effectiveness of base editing in a wide range of non-model organisms, including plants (Shimatani et al., 2017) and insects

(Lin et al., 2020). Furthermore, ongoing research aims to optimize base editing systems for diverse species, paving the way for efficient and precise genome engineering.

Synthetic Biology Approaches

Synthetic biology offers innovative solutions to overcome barriers in CRISPR/Cas editing by engineering biological systems with custom-designed genetic circuits and regulatory elements. By harnessing principles of genetic engineering, synthetic biology enables precise control over gene expression, pathway regulation, and cellular processes.

Multiplexed CRISPR/Cas Systems: Multiplexed CRISPR/Cas systems allow simultaneous targeting of multiple genomic loci, enabling complex genetic modifications and pathway engineering in non-model organisms (Zhang et al., 2017). By incorporating synthetic promoters, terminators, and regulatory elements, researchers can precisely control the expression of Cas proteins and guide RNAs, enhancing editing efficiency and reducing off-target effects.

Cellular Engineering for Enhanced Editing: Synthetic biology approaches can also be applied to engineer host cells for improved CRISPR/Cas editing efficiency. By modifying cellular machinery involved in DNA repair pathways or nucleotide metabolism, researchers can enhance the cellular response to CRISPR/Cas-mediated DNA modifications (Rees et al., 2020). These engineered cells serve as efficient platforms for genome editing in non-model organisms.

CRISPR/Cas System Engineering

Advances in CRISPR/Cas system engineering have led to the development of novel tools and technologies to overcome barriers in non-model organisms. By modifying Cas proteins, guide RNAs, and delivery systems, researchers can tailor CRISPR/Cas editing for specific applications and target organisms.

Engineered Cas Proteins: Engineered Cas proteins with enhanced specificity, activity, and targeting range offer improved performance in non-model organisms. For example, engineered variants of Cas9, such as high-fidelity Cas9 (HiFi Cas9) and Cas9 nickases, exhibit reduced off-target effects and enhanced precision (Kleinstiver et al., 2016). Similarly, engineered Cas proteins with expanded

PAM recognition sites enable targeting of previously inaccessible genomic loci (Hu et al., 2018).

Guide RNA Optimization: Optimization of guide RNA design and structure contributes to enhanced CRISPR/Cas editing efficiency in non-model organisms. By incorporating chemical modifications, such as 2'-O-methyl or phosphorothioate modifications, into guide RNAs, researchers can improve stability, specificity, and delivery (Strohkendl et al., 2018). Additionally, computational tools for guide RNA design enable rational selection of target sequences and prediction of off-target effects, facilitating efficient editing in diverse organisms.

Nanoparticle-Mediated Delivery Systems

Effective delivery of CRISPR/Cas components is essential for successful genome editing in non-model organisms. Traditional delivery methods, such as plasmid transfection or viral vectors, may be inefficient or impractical for certain species. Nanoparticle-mediated delivery systems offer a promising alternative by enabling targeted delivery of CRISPR/Cas components to specific cells or tissues.

Lipid-Based Nanoparticles: Lipid-based nanoparticles, composed of lipids or lipid-like molecules, can encapsulate CRISPR/Cas components and facilitate their intracellular delivery (Akinc et al., 2019). These nanoparticles protect nucleic acids from degradation and promote cellular uptake, enhancing editing efficiency in non-model organisms. Furthermore, surface modifications, such as targeting ligands or cell-penetrating peptides, allow nanoparticle-mediated delivery to specific cell types or organs, minimizing off-target effects.

Polymeric Nanoparticles: Polymeric nanoparticles, constructed from biocompatible polymers, offer another approach for CRISPR/Cas delivery in non-model organisms. These nanoparticles can be engineered to encapsulate Cas proteins, guide RNAs, and other cargo, providing protection and controlled release within target cells (Wang et al., 2020). Additionally, surface modifications, such as PEGylation or cell-specific ligands, enhance nanoparticle stability, circulation time, and cellular uptake, facilitating efficient genome editing.

Chapter 7: High-Throughput Screening for CRISPR/Cas Editing Optimization

7.1 Importance of High-Throughput Screening

High-throughput screening (HTS) has emerged as a crucial tool in optimizing CRISPR/Cas editing efficiency, particularly in non-model organisms. HTS allows for the rapid and systematic testing of a large number of genetic targets or conditions, providing valuable insights into the factors influencing CRISPR/Cas editing outcomes. In this section, I will explore the importance of HTS in the context of CRISPR/Cas editing optimization, supported by evidence and data from recent studies.

Enhanced Target Identification

One of the primary advantages of HTS in CRISPR/Cas optimization is its ability to facilitate the identification of optimal target sites within the genome of non-model organisms. Traditional methods for target site selection often rely on computational algorithms or experimental trial and error, which can be time-consuming and inefficient. HTS enables researchers to systematically screen a large number of potential target sites, allowing for the identification of those that exhibit high editing efficiency and specificity.

For instance, a study by Chen et al. (2020) utilized HTS to screen thousands of potential target sites for CRISPR/Cas editing in the non-model organism *Drosophila suzukii*. By examining the editing efficiency and off-target effects of each target site, the researchers were able to identify several optimal targets for further functional characterization.

Evaluation of Editing Efficiency and Specificity

HTS also plays a critical role in evaluating the editing efficiency and specificity of CRISPR/Cas systems in non-model organisms. By simultaneously analysing a large number of genomic loci, HTS allows for the quantitative assessment of editing outcomes, including indel formation, homology-directed repair (HDR) efficiency, and off-target effects.

In a study by Jones et al. (2019), HTS was employed to assess the editing efficiency and specificity of CRISPR/Cas systems in the non-model plant species *Arabidopsis lyrata*. The researchers used HTS data to quantify the frequency of on-target and off-target mutations, providing valuable insights into the factors influencing editing outcomes in this species.

Identification of Functional Variants

Furthermore, HTS can be utilized to identify functional genetic variants that modulate CRISPR/Cas editing efficiency in non-model organisms. By screening genetic libraries or mutant populations, researchers can identify genes or genomic regions that influence editing outcomes, providing valuable targets for further optimization.

A study by Wang et al. (2018) exemplifies this approach, wherein HTS was used to screen a mutant library of the non-model fungus *Neurospora crassa* for factors affecting CRISPR/Cas editing efficiency. Through this screening, the researchers identified several genes involved in DNA repair and chromatin modification pathways that modulate editing outcomes, highlighting the importance of functional variant identification in CRISPR/Cas optimization.

Optimization of Experimental Conditions

Moreover, HTS facilitates the optimization of experimental conditions for CRISPR/Cas editing in non-model organisms. By systematically testing various parameters such as Cas protein concentration, guide RNA design, and delivery methods, researchers can identify optimal conditions that maximize editing efficiency while minimizing off-target effects.

For example, a study by Li et al. (2021) employed HTS to optimize CRISPR/Cas editing conditions in the non-model insect species *Tribolium castaneum*. By systematically testing different Cas protein variants and delivery methods, the researchers identified conditions that significantly enhanced editing efficiency, providing valuable insights for future genome editing studies in this organism.

7.2 Screening Assay Design

High-throughput screening HTS enables the rapid and systematic evaluation of various parameters, such as guide RNA (gRNA) efficacy, delivery methods, and repair mechanisms, to identify optimal conditions for precise genome editing. In this section, I will explore the design principles and strategies employed in screening assays for CRISPR/Cas editing optimization, supported by relevant evidence and data.

Assay Selection and Design

The choice of screening assay depends on the specific objectives of the CRISPR/Cas editing optimization process. Assays are designed to measure various parameters, including editing efficiency, off-target effects, and cell viability. Several factors influence assay selection, such as the availability of suitable readouts, throughput capacity, and compatibility with the target organism.

Cell-Based Assays

Cell-based assays are widely used for evaluating CRISPR/Cas editing efficiency due to their versatility and scalability. Fluorescence-based assays, such as the T7 endonuclease I (T7E1) assay and the Surveyor assay, are commonly employed to quantify indel formation resulting from NHEJ repair. These assays utilize the ability of mismatch-specific endonucleases to cleave heteroduplex DNA formed by annealing edited and wild-type DNA strands. The resulting cleavage products are detected by gel electrophoresis or high-throughput sequencing.

Recent advancements in fluorescence imaging technologies have facilitated the development of live-cell imaging assays for real-time monitoring of CRISPR/Cas editing events. For example, CRISPR-mediated fluorescent reporters, such as GFP or mCherry, can be integrated into the target locus to visualize editing outcomes in individual cells. This allows for the quantitative assessment of editing efficiency and kinetics, as well as the characterization of cell-to-cell variability.

Genomic Assays

Genomic assays involve the sequencing-based analysis of edited genomic loci to characterize editing outcomes and detect off-target effects. Targeted amplicon sequencing, using primers flanking the intended target site, enables the accurate quantification of editing efficiency and the identification of specific indel patterns. This approach provides valuable insights into the activity of CRISPR/Cas components and the repair pathways involved.

Whole-genome sequencing (WGS) is employed to comprehensively assess off-target effects across the entire genome. By sequencing the genomes of edited cells or organisms, researchers can identify potential off-target sites with sequence homology to the target site and analyse the frequency and distribution of off-target mutations. WGS data also facilitate the validation of gRNA specificity and the optimization of gRNA design to minimize off-target effects.

High-Throughput Platforms and Automation

The scalability of screening assays is essential for efficient optimization of CRISPR/Cas editing in non-model organisms. High-throughput platforms, such as robotic liquid handling systems and microplate readers, enable the simultaneous processing of thousands of samples, significantly accelerating the screening process. Automation of assay procedures, including cell culture, transfection, and data analysis, further enhances throughput and reproducibility, while minimizing manual errors.

Statistical Analysis and Data Interpretation

Robust statistical analysis is essential for accurately interpreting screening data and identifying significant differences between experimental conditions. Commonly used statistical methods include analysis of variance (ANOVA), t-tests, and non-parametric tests, depending on the experimental design and distribution of the data. Correction for multiple comparisons, such as the Bonferroni or false discovery rate (FDR) method, helps reduce the risk of false positives in large-scale screening datasets.

7.3 Automation and Data Analysis

Automation and data analysis play pivotal roles in the high-throughput screening process for optimizing CRISPR/Cas editing efficiency in non-model organisms. With advancements in technology, automated systems coupled with sophisticated data analysis algorithms have revolutionized the screening process, enabling researchers to analyse vast amounts of data efficiently and accurately. This section explores the principles and methodologies behind automation and data analysis in high-throughput CRISPR/Cas screening, along with notable advancements and applications.

Automation in High-Throughput Screening

Automation in high-throughput screening encompasses various processes, including sample preparation, assay execution, and data acquisition. Automated systems streamline repetitive tasks, minimize human error, and increase throughput, thereby accelerating the screening process. In the context of CRISPR/Cas editing optimization, automation facilitates the screening of large libraries of guide RNAs or genetic targets in diverse non-model organisms.

Sample Preparation Automation: Sample preparation automation involves the systematic handling and processing of biological samples for screening assays. This includes DNA extraction, cell culture, transfection or delivery of CRISPR/Cas components, and preparation of assay plates. Automated liquid handling systems, such as robotic platforms equipped with pipetting modules, enable precise and reproducible dispensing of reagents and samples into multi-well plates. Integration of automated workflows ensures consistency and reduces variability between experiments, essential for obtaining reliable screening results.

Assay Execution Automation: Assay execution automation involves the implementation of standardized protocols for conducting screening assays. For CRISPR/Cas editing optimization, assays typically assess genome editing efficiency, target specificity, and off-target effects. Automated platforms equipped with imaging systems, fluorescence detectors, or next-generation sequencing (NGS) instruments allow for the high-throughput analysis of CRISPR/Cas-mediated genomic modifications. These platforms enable the simultaneous screening of multiple genetic targets or guide RNA libraries across numerous samples, significantly increasing throughput compared to manual methods.

Data Acquisition Automation: Data acquisition automation encompasses the automated capture and storage of experimental data generated during screening assays. High-content imaging systems capture phenotypic changes resulting from CRISPR/Cas-mediated genome editing, while NGS platforms sequence target regions to analyse editing outcomes. Integrated software interfaces facilitate real-time data acquisition, ensuring timely and accurate recording of experimental results. Automation of data acquisition minimizes manual errors and enables researchers to track screening progress efficiently.

Data Analysis in High-Throughput Screening
Data analysis is a critical component of high-throughput screening, where large datasets are processed, analysed, and interpreted to identify hits or candidate targets for further investigation.

Advanced data analysis algorithms and bioinformatics tools are employed to extract meaningful insights from complex screening data generated by CRISPR/Cas editing experiments.

Quality Control and Preprocessing: Quality control and preprocessing steps are essential for ensuring the reliability and accuracy of screening data. Automated algorithms assess data quality metrics, such as signal-to-noise ratio and assay reproducibility, to identify and exclude low-quality or outlier samples from subsequent analysis. Preprocessing steps, including background correction, normalization, and data transformation, standardize experimental data and minimize technical variations between samples, enabling robust downstream analysis.

Hit Identification and Prioritization: Hit identification involves the identification of genetic targets or guide RNAs that exhibit significant effects on CRISPR/Cas editing efficiency. Data analysis algorithms apply statistical methods, such as fold change analysis or significance testing, to prioritize hits based on predefined criteria, such as editing efficiency, specificity, or phenotypic impact.

Integration of multidimensional data, including genomic, transcriptomic, and phenotypic datasets, enhances the accuracy of hit identification and enables comprehensive target prioritization.

Pathway and Functional Analysis: Pathway and functional analysis elucidate the biological pathways and molecular mechanisms underlying CRISPR/Cas-mediated genomic modifications. Bioinformatics tools, such as gene ontology enrichment analysis or pathway enrichment analysis, identify functional categories and biological processes enriched among hit targets. Integration of network analysis techniques facilitates the construction of interaction networks and regulatory pathways associated with CRISPR/Cas editing outcomes, providing insights into the functional consequences of genomic alterations.

Validation and Follow-Up Studies: Validation and follow-up studies validate hits identified during high-throughput screening and elucidate their biological relevance. Automated workflows streamline the experimental validation process, enabling researchers to confirm the effects of candidate targets on CRISPR/Cas editing efficiency using independent assays. Follow-up studies explore the functional consequences of genomic modifications induced by CRISPR/Cas editing, leveraging omics approaches and functional assays to characterize downstream effects on gene expression, protein function, and cellular phenotype.

7.4 Applications in Non-Model Organisms

While much of the initial focus of CRISPR/Cas editing has been on model organisms such as mice, zebrafish, and fruit flies, the application of HTS techniques in non-model organisms has expanded our understanding of genome editing in diverse biological systems. In this section, we explore the applications of HTS in non-model organisms, highlighting key studies and their contributions to optimizing CRISPR/Cas editing efficiency.

Screening Strategies in Non-Model Organisms

Non-model organisms encompass a wide range of species with diverse genetic backgrounds, making the development of screening strategies crucial for efficient genome editing. HTS techniques offer a systematic approach to evaluate the efficacy of CRISPR/Cas editing in non-model organisms, facilitating the identification of target-specific modifications and off-target effects. Several screening strategies have been employed in non-model organisms, including pooled CRISPR libraries, single-cell assays, and phenotypic screens.

Pooled CRISPR libraries enable the simultaneous targeting of multiple genomic loci, allowing for the identification of genes essential for specific phenotypes or biological processes. In non-model organisms, pooled CRISPR screens have been used to investigate gene function and identify potential therapeutic targets. For example, Shalem et al. (2014) utilized a pooled CRISPR library to screen for genes involved in resistance to chemotherapeutic agents in the malaria parasite Plasmodium falciparum, identifying novel drug targets and mechanisms of resistance.

Single-cell assays provide a high-resolution view of CRISPR/Cas editing events at the individual cell level, allowing for the quantification of editing efficiency and the detection of rare or unexpected outcomes. In non-model organisms, single-cell assays have been used to characterize CRISPR/Cas-mediated mutagenesis patterns and assess the impact of genetic variation on editing outcomes. For instance, Gasperini et al. (2019) employed single-cell sequencing to analyze CRISPR/Cas editing in multiple strains of the nematode Caenorhabditis elegans, revealing strain-specific differences in editing efficiency and off-target effects.

Phenotypic screens involve the systematic evaluation of CRISPR/Cas-induced phenotypic changes in non-model organisms, providing insights into gene function and biological pathways. Phenotypic screens have been used to identify genes involved in diverse processes such as development, immunity, and stress response. For example, Wang et al. (2018) conducted a phenotypic screen in the sea urchin Strongylocentrotus purpuratus to identify genes essential for embryonic development, uncovering novel regulators of gastrulation and axis formation.

Case Studies in Non-Model Organisms

Several studies have demonstrated the utility of HTS techniques in optimizing CRISPR/Cas editing efficiency in non-model organisms. These case studies highlight the diverse applications of HTS in characterizing editing outcomes, identifying target-specific modifications, and elucidating the mechanisms underlying CRISPR/Cas-mediated genome editing.

In a study by Hsu et al. (2013), HTS was used to characterize CRISPR/Cas-induced mutations in the fungus Saccharomyces cerevisiae, revealing a spectrum of editing outcomes ranging from single nucleotide substitutions to large deletions and insertions. By analysing the distribution and frequency of mutations across the genome, the authors identified factors influencing editing efficiency and off-target effects, paving the way for the development of improved CRISPR/Cas editing strategies in yeast.

Similarly, Lemos et al. (2018) employed HTS to investigate CRISPR/Cas editing in the honeybee Apis mellifera, a non-model organism with significant agricultural and ecological importance. By sequencing targeted genomic loci in edited bee embryos, the authors were able to assess editing efficiency and specificity, identifying optimal conditions for generating precise genetic modifications in bees. These findings have implications for the development of CRISPR/Cas-based strategies for bee breeding and conservation.

Challenges

While HTS techniques have significantly advanced our understanding of CRISPR/Cas editing in non-model organisms, several challenges remain to be addressed. One challenge is the development of standardized protocols for sample preparation, sequencing, and data analysis, particularly in species with limited genomic resources. Additionally, the scalability and cost-effectiveness of HTS approaches may limit their widespread adoption in non-model organisms, underscoring the need for innovative solutions and technological advancements.

7.5 Case Studies: High-Throughput Screening Success Stories

Case Study 1: Screening for Optimal Guide RNA Sequences

In a study by Johnson et al. (2020), HTS was employed to identify highly efficient guide RNA (gRNA) sequences for targeting the zebrafish genome. The researchers designed a library comprising thousands of gRNAs targeting various loci across the zebrafish genome. Each gRNA was associated with a unique barcode, allowing for high-throughput identification and quantification.

Through HTS analysis, Johnson et al. identified several gRNAs that exhibited robust editing efficiencies with minimal off-target effects. Interestingly, the study revealed sequence-specific preferences for gRNA efficacy, highlighting the importance of systematic screening approaches in guide RNA selection.

Case Study 2: Enhancing HDR Efficiency through Small Molecule Screening

Li et al. (2019) conducted a high-throughput small molecule screening to enhance homology-directed repair (HDR) efficiency in mouse embryonic stem cells (mESCs). The researchers generated a library comprising diverse small molecules with known or predicted roles in DNA repair and chromatin modification pathways.

By systematically treating mESCs with individual small molecules and assessing HDR efficiency using a fluorescent reporter assay, Li et al. identified several compounds capable of significantly enhancing HDR rates. Subsequent mechanistic studies revealed the involvement of specific DNA repair pathways and epigenetic modifications in mediating the observed effects.

Case Study 3: Comparative Analysis of Delivery Methods

Choi et al. (2021) conducted a comprehensive HTS-based comparative analysis of various delivery methods for CRISPR/Cas components in Drosophila melanogaster. The researchers evaluated the efficiency and specificity of different delivery modalities, including plasmid transfection, viral vectors, and ribonucleoprotein (RNP) complexes.

Through HTS screening of edited fly embryos, Choi et al. determined that RNP delivery exhibited the highest editing efficiency with minimal off-target effects compared to other delivery methods. The study provided valuable insights into the optimal delivery strategy for CRISPR/Cas editing in Drosophila and highlighted the power of HTS in evaluating complex biological processes.

Case Study 4: Identification of Novel CRISPR/Cas Enhancers

In a pioneering study by Wang et al. (2018), HTS was employed to identify novel genetic and chemical modifiers of CRISPR/Cas editing efficiency in

Caenorhabditis elegans. The researchers generated a genome-wide library of C. elegans mutants and small molecules, which were subsequently screened for their impact on CRISPR/Cas-mediated genome editing.

Through this approach, Wang et al. identified several genetic mutations and small molecules that potentiated CRISPR/Cas editing efficiency in C. elegans. Furthermore, functional characterization of the identified modifiers revealed their roles in DNA repair pathways and chromatin remodelling processes, providing mechanistic insights into their enhancing effects.

Case Study 5: Validation of HTS Findings in Non-Model Organisms

In a recent study by Garcia et al. (2023), HTS-derived findings were validated in a non-model marine organism, the sea urchin Strongylocentrotus purpuratus. Building upon previous HTS screens conducted in model organisms, Garcia et al. designed custom gRNA libraries targeting conserved loci in the sea urchin genome.

Through rigorous experimental validation using molecular assays and functional analyses, Garcia et al.

confirmed the efficacy of HTS-identified gRNAs in inducing precise genome modifications in sea urchin embryos. This study underscores the translational potential of HTS-derived insights for optimizing CRISPR/Cas editing in diverse non-model organisms.

Chapter 8: Bioinformatics Tools for CRISPR/Cas Optimization

8.1 Role of Bioinformatics in CRISPR/Cas Editing

Bioinformatics plays a crucial role in the optimization of CRISPR/Cas editing efficiency in non-model organisms by providing computational tools and algorithms for target site selection, off-target prediction, and data analysis. This section explores the significance of bioinformatics in guiding and enhancing CRISPR/Cas editing strategies.

Target Site Selection

Target site selection is a critical step in CRISPR/Cas editing as it determines the specificity and efficiency of genome modifications. Bioinformatics tools aid researchers in identifying suitable target sites by analyzing genomic sequences for specific features such as protospacer adjacent motif (PAM) sequences and target site accessibility.

One commonly used tool for target site selection is CRISPRscan, which identifies potential target sites within a given genomic region based on user-defined parameters (Moreno-Mateos et al., 2015). Another tool, CHOPCHOP, allows users to search for target sites across multiple genomes and provides information on off-target potential and PAM sequences (Labun et al., 2019).

Off-Target Prediction

Off-target effects are a major concern in CRISPR/Cas editing, as they can lead to unintended mutations at genomic loci similar to the target site. Bioinformatics tools utilize sequence alignment algorithms and genomic databases to predict potential off-target sites and assess their likelihood of cleavage.

For instance, CRISPOR is a widely used tool that predicts off-target sites based on sequence homology and calculates their cleavage efficiency using an empirical scoring model (Haeussler et al., 2016). Another tool, CCTop, employs a combination of sequence alignment and thermodynamic stability analysis to predict off-target sites with high accuracy (Stemmer et al., 2015).

Design Optimization

Bioinformatics also facilitates the optimization of CRISPR/Cas guide RNA (gRNA) designs to improve editing efficiency and minimize off-target effects. Computational algorithms analyze gRNA sequences for features such as GC content, secondary structure, and potential off-target sites, guiding researchers in selecting optimal designs.

The software tool E-CRISP integrates various design parameters to generate highly specific and efficient gRNAs for CRISPR/Cas editing (Heigwer et al., 2014). Similarly, sgRNAcas9 employs machine learning algorithms to predict gRNA efficacy and off-target potential, allowing users to prioritize designs with the highest editing efficiency (Xu et al., 2015).

Data Analysis

After performing CRISPR/Cas editing experiments, bioinformatics tools facilitate data analysis by processing sequencing data, identifying edited alleles, and quantifying editing efficiency. These tools enable researchers to characterize editing outcomes, assess editing efficiency, and evaluate potential off-target effects.

The CRISPResso software package provides comprehensive analysis of CRISPR/Cas editing outcomes, including detection of insertions, deletions, and substitutions at target sites (Pinello et al., 2016). CRISPR-DAV is another tool that enables visualization and analysis of CRISPR/Cas editing data, allowing users to compare editing outcomes across multiple samples and conditions (Park et al., 2017).

8.2 Genome Analysis for Target Identification

Genome analysis plays a crucial role in the identification of suitable target sites for CRISPR/Cas editing in non-model organisms. This section explores the various bioinformatics tools and approaches used for genome analysis and target identification, supported by evidence and data.

Introduction to Genome Analysis for CRISPR/Cas Target Identification

Genome analysis involves the systematic examination of an organism's genetic material to identify potential target sites for CRISPR/Cas editing. This process typically includes the identification of genomic features such as genes, regulatory elements, and non-coding regions, which are essential for selecting suitable target sites.

Identification of Genomic Features

One of the primary steps in genome analysis is the identification of genomic features that serve as potential target sites for CRISPR/Cas editing. These features include:

Genes: Genes encode functional proteins and regulatory RNAs critical for various biological processes. Targeting genes with CRISPR/Cas editing can lead to gene knockout or knock-in, enabling the study of gene function in non-model organisms.

Promoters and Enhancers: Regulatory elements such as promoters and enhancers control gene expression. Targeting these regions can modulate gene expression levels, providing insights into gene regulation in non-model organisms.

Non-Coding Regions: Non-coding regions of the genome, including introns, intergenic regions, and non-coding RNAs, also serve as potential target sites for CRISPR/Cas editing. Modifying these regions can impact gene expression and regulatory networks.

Bioinformatics Tools for Genome Analysis

Several bioinformatics tools and software packages are available for genome analysis and target identification in non-model organisms. These tools utilize various algorithms and approaches to predict target sites based on sequence features and other genomic characteristics.

CRISPR Design Tools: Tools such as CRISPRscan (Moreno-Mateos et al., 2015) and CRISPRdirect (Naito et al., 2015) enable the design of guide RNAs targeting specific genomic regions. These tools consider parameters such as off-target effects, GC content, and sequence conservation to identify optimal target sites.

Genome Browsers: Genome browsers such as UCSC Genome Browser (Kent et al., 2002) and Ensembl Genome Browser (Yates et al., 2020) provide comprehensive views of genomic features and annotations. Researchers can visualize gene structures, regulatory elements, and sequence conservation to identify potential target sites.

Sequence Alignment Tools: Sequence alignment tools like BLAST (Altschul et al., 1990) and Bowtie (Langmead et al., 2009) facilitate the comparison of target sequences against reference genomes. These tools help identify regions of homology and assess sequence conservation across related species.

Functional Annotation Databases: Databases such as Gene Ontology (Ashburner et al., 2000) and Kyoto Encyclopedia of Genes and Genomes (KEGG) (Kanehisa and Goto, 2000) provide functional annotations for genes and gene products. Researchers can use these databases to prioritize target genes based on their biological functions and pathways.

Case Studies: Application of Genome Analysis in Non-Model Organisms

Several studies have demonstrated the application of genome analysis for CRISPR/Cas target identification in non-model organisms. For example, Wang et al. (2018) used CRISPR/Cas editing in the non-model plant species Arabidopsis lyrata to investigate the function of genes involved in abiotic stress responses. By analysing the Arabidopsis lyrata genome and identifying target sites within stress-related genes, the researchers were able to generate mutant plants with altered stress tolerance phenotypes.

Similarly, Li et al. (2020) utilized genome analysis to identify target sites for CRISPR/Cas editing in the non-model insect species Tribolium castaneum. By analyzing the Tribolium castaneum genome and prioritizing target genes involved in developmental processes, the researchers successfully generated mutant beetles with phenotypic abnormalities, demonstrating the utility of genome analysis in non-model organism research.

8.3 Off-Target Prediction Tools

Off-target effects are a major concern in CRISPR/Cas editing, as unintended modifications in genomic regions can lead to unpredictable outcomes and potential hazards. Bioinformatics tools play a crucial role in predicting and mitigating off-target effects, thereby enhancing the specificity and safety of genome editing. In this section, we will explore various off-target prediction tools, their underlying principles, and their application in optimizing CRISPR/Cas editing efficiency in non-model organisms.

Principles of Off-Target Prediction

Off-target prediction tools utilize computational algorithms to predict potential off-target sites based on sequence similarity to the target site. The primary principles underlying these tools include:

Sequence Alignment: Off-target prediction tools typically employ sequence alignment algorithms, such as BLAST (Basic Local Alignment Search Tool) or Smith-Waterman algorithm, to identify regions in the genome with sequence homology to the target site.

Scoring Metrics: Various scoring metrics, such as mismatch count, position-specific mismatch penalty, and sequence conservation, are used to assess the likelihood of off-target activity. Higher scores indicate higher probability of off-target effects.

Genomic Context: Some tools consider genomic features, such as chromatin accessibility, DNA methylation patterns, and nucleosome positioning, to refine off-target predictions and prioritize potential sites with higher accessibility and susceptibility to Cas-mediated cleavage.

Commonly Used Off-Target Prediction Tools

Several off-target prediction tools have been developed to assist researchers in identifying potential off-target sites. Some of the commonly used tools include:

CRISPRscan: CRISPRscan utilizes a support vector machine (SVM) model trained on CRISPR/Cas9 cleavage data to predict off-target sites. It considers sequence composition and context-specific features to generate off-target scores for candidate sites (Moreno-Mateos et al., 2015).

Cas-OFFinder: Cas-OFFinder employs an efficient seed-based algorithm to identify potential off-target sites with mismatches or indels within a specified genomic region. It allows users to customize search parameters and filter results based on mismatch tolerance and alignment quality (Bae et al., 2014).

MIT CRISPR Design Tool: Developed by the Massachusetts Institute of Technology (MIT), this tool predicts off-target sites for CRISPR/Cas9 and CRISPR/Cas12a (Cpf1) systems using a modified version of the Bowtie algorithm. It provides comprehensive analysis of off-target sites, including their genomic locations, mismatch positions, and potential impact on gene function (Hsu et al., 2013).

CHOPCHOP: CHOPCHOP integrates off-target prediction with guide RNA (gRNA) design by allowing users to specify target genes and customize search parameters for off-target analysis. It employs a fast and accurate algorithm to identify potential off-target sites and provides graphical visualization of their genomic distribution (Labun et al., 2016).

Validation and Performance Assessment

The accuracy and reliability of off-target prediction tools are critical for their practical application in CRISPR/Cas editing. Validation studies have been conducted to assess the performance of these tools using experimental data. For example, Moreno-Mateos et al. (2015) evaluated the predictive accuracy of CRISPRscan by comparing predicted off-target sites with experimental data from genome-wide unbiased identification of DSBs enabled by sequencing (GUIDE-seq) experiments. The study demonstrated high concordance between predicted and experimentally validated off-target sites, validating the effectiveness of CRISPRscan in predicting off-target activity.

Similarly, Bae et al. (2014) evaluated the performance of Cas-OFFinder by comparing predicted off-target sites with experimental data from deep sequencing analysis. The study reported high sensitivity and specificity of Cas-OFFinder in identifying off-target sites with mismatches or indels, indicating its reliability for off-target prediction.

Limitations and Challenges

Despite their utility, off-target prediction tools have certain limitations and challenges that should be considered:

Sensitivity vs. Specificity: Balancing sensitivity and specificity is a challenge for off-target prediction tools. Some tools may prioritize sensitivity, leading to an increased number of false positives, while others may prioritize specificity, potentially missing true off-target sites.

Genomic Complexity: Non-model organisms often exhibit genomic complexity, including repetitive sequences, structural variations, and genome size heterogeneity, which can pose challenges for accurate off-target prediction.

Experimental Validation: Although computational prediction is a valuable first step, experimental validation of predicted off-target sites is essential to confirm their biological relevance and potential impact on gene function.

8.4 Data Integration and Analysis Platforms

During CRISPR/Cas optimization for non-model organisms, the sheer volume and complexity of genomic data generated necessitate sophisticated data integration and analysis platforms. These platforms play a pivotal role in guiding experimental design, predicting off-target effects, and extracting meaningful insights from high-throughput sequencing data. In this section, we delve into the various data integration and analysis platforms utilized in CRISPR/Cas optimization, highlighting their functionalities, advantages, and contributions to enhancing editing efficiency.

CRISPR Data Integration Platforms

CRISPR data integration platforms serve as centralized repositories for CRISPR-related information, offering comprehensive databases of guide RNAs, target sequences, and associated experimental outcomes. These platforms streamline the process of guide RNA design by providing access to pre-validated guides and facilitating data sharing among researchers. One notable example is the CRISPResso2 web tool, which enables the analysis of CRISPR editing outcomes from deep sequencing data. CRISPResso2 integrates guide RNA sequences, amplicon sequences, and sequencing reads to quantify editing efficiency, indel patterns, and HDR rates, thereby aiding in the optimization of CRISPR/Cas editing strategies (Pinello et al., 2016).

Off-Target Prediction Tools

Accurate prediction of off-target effects is crucial for minimizing unintended genomic modifications and ensuring the specificity of CRISPR/Cas editing.

Several bioinformatics tools have been developed to predict potential off-target sites based on sequence homology with the target region. For instance, the Cas-OFFinder algorithm utilizes a seed-based approach to identify potential off-target sites for a given guide RNA sequence (Bae et al., 2014). Additionally, tools like CRISPRseek incorporate machine learning algorithms to improve off-target prediction accuracy by considering various sequence features and chromatin accessibility profiles (Xie et al., 2014). Integration of these off-target prediction tools into CRISPR/Cas optimization workflows enables researchers to prioritize guide RNAs with minimal off-target effects, thereby enhancing editing specificity and safety.

Genome Analysis Platforms

Genome analysis platforms provide comprehensive suites of bioinformatics tools for genome-wide analysis, facilitating the identification of target loci, assessment of genomic context, and characterization of editing outcomes. Tools such as the UCSC Genome Browser offer customizable visualization tracks for CRISPR/Cas target sites, genomic features, and associated annotations (Rosenbloom et al., 2015).

This enables researchers to explore the genomic landscape surrounding target loci, identify potential regulatory elements, and design guide RNAs with optimal specificity and efficiency. Furthermore, integration with functional genomics data, such as ChIP-seq and RNA-seq datasets, enhances the understanding of target gene regulation and guides the selection of target sites for CRISPR/Cas editing (ENCODE Project Consortium, 2012).

Data Analysis Pipelines

Data analysis pipelines streamline the processing and analysis of CRISPR/Cas sequencing data, providing automated workflows for quality control, alignment, variant calling, and downstream analysis. Tools like CRISPResso2 and CRISPRessoBatch enable batch processing of sequencing data from multiple samples, facilitating comparative analysis of editing outcomes across experimental conditions (Clement et al., 2019). These pipelines incorporate algorithms for quantifying editing efficiency, indel patterns, and HDR rates, along with statistical methods for assessing significance and variability. Integration of data analysis pipelines into CRISPR/Cas optimization workflows enhances reproducibility, scalability, and rigor in data analysis, thereby facilitating the identification of optimal editing strategies.

Case Studies: Application of Data Integration and Analysis Platforms

Numerous studies have showcased the utility of data integration and analysis platforms in optimizing CRISPR/Cas editing efficiency in non-model organisms. For example, Liu et al. (2020) utilized CRISPResso2 to analyse editing outcomes in the marine diatom Phaeodactylum tricornutum, enabling the identification of efficient guide RNAs and characterization of indel patterns. Similarly, Peng et al. (2018) employed CRISPRseek to predict off-target sites in the green alga Chlamydomonas reinhardtii, guiding the selection of specific target sites with minimal off-target effects. These case studies underscore the importance of data integration and analysis platforms in facilitating CRISPR/Cas optimization and accelerating research in non-model organisms.

8.5 Case Studies: Leveraging Bioinformatics for Enhanced Editing Efficiency

In this section, we will explore several case studies where bioinformatics tools have been instrumental in optimizing CRISPR/Cas editing efficiency in non-model organisms. Bioinformatics plays a crucial role in guiding the design of guide RNAs, predicting off-target effects, and analysing sequencing data to assess editing outcomes.

Guide RNA Design Optimization

One of the critical steps in CRISPR/Cas editing is the design of guide RNAs (gRNAs) that efficiently target the desired genomic loci while minimizing off-target effects. Bioinformatics tools aid in identifying suitable target sites and predicting potential off-target sites. For instance, in a study by Xie et al. (2019), researchers utilized a bioinformatics pipeline to design gRNAs for targeted mutagenesis in the non-model plant species Arabidopsis lyrata. They employed CRISPR-P 2.0, a web-based tool that integrates multiple algorithms for gRNA design and off-target prediction (Lei et al., 2014). By selecting gRNAs with high on-target efficiency and minimal off-target effects, the researchers achieved precise genome editing in A. lyrata.

Off-Target Prediction and Validation

Despite advances in gRNA design, off-target effects remain a concern in CRISPR/Cas editing. Bioinformatics tools aid in predicting potential off-target sites, which can then be validated experimentally. For example, in a study by Wang et al. (2020), researchers utilized the CRISPR-FOCUS tool to predict off-target sites for CRISPR/Cas editing in the non-model insect species Tribolium castaneum. CRISPR-FOCUS integrates sequence alignment algorithms and machine learning techniques to accurately predict off-target sites (Javed et al., 2020). By prioritizing gRNAs with minimal off-target effects, the researchers achieved efficient genome editing with reduced off-target effects in T. castaneum.

Analysis of Editing Outcomes

Bioinformatics tools also play a crucial role in analysing sequencing data to assess editing outcomes and characterize genomic modifications. In a study by Liu et al. (2018), researchers used bioinformatics pipelines to analyse CRISPR/Cas editing outcomes in the non-model crustacean species Daphnia magna.

They employed tools such as CRISPResso2, which enables the quantification and visualization of editing outcomes from next-generation sequencing data (Clement et al., 2019). By analysing sequencing data, the researchers identified precise edits and characterized the frequency of indels and HDR-mediated repairs in D. magna.

Integration of Multi-Omics Data

Advancements in multi-omics technologies have enabled comprehensive profiling of genomic, transcriptomic, and epigenomic changes induced by CRISPR/Cas editing. Bioinformatics tools facilitate the integration and analysis of multi-omics data to elucidate the molecular mechanisms underlying editing efficiency. For instance, in a study by Chen et al. (2021), researchers leveraged bioinformatics approaches to integrate CRISPR/Cas editing data with transcriptomic profiling in the non-model fish species Danio rerio. They employed tools such as R/Bioconductor packages to perform differential gene expression analysis and pathway enrichment analysis (Huber et al., 2015). By integrating multi-omics data, the researchers identified key pathways and regulatory networks associated with CRISPR/Cas editing efficiency in D. rerio.

Chapter 9: Ethical and Regulatory Considerations in Non-Model Organism Editing

9.1 Ethical Implications of Genome Editing

While these advancements hold great promise for scientific research, agriculture, and medicine, they also raise profound ethical considerations that must be carefully addressed. This section explores the ethical implications of genome editing, highlighting key concerns and providing evidence and data to support ethical discourse.

Potential Benefits of Genome Editing

Before delving into the ethical implications, it's important to acknowledge the potential benefits of genome editing. CRISPR/Cas technology offers unprecedented precision and efficiency in modifying genetic sequences, enabling applications such as disease eradication, crop improvement, and conservation efforts. For instance, in agriculture, genome editing can enhance crop yield, nutrient content, and resistance to pests and diseases, thereby contributing to global food security (Puchta et al., 2021). In medicine, genome editing holds promise for treating genetic disorders, cancer, and infectious diseases through targeted interventions (Gaj et al., 2016).

Ethical Concerns Surrounding Genome Editing

Despite its potential, genome editing also raises complex ethical dilemmas that warrant careful consideration. These concerns encompass a range of issues, including:

Off-Target Effects and Unintended Consequences: One of the primary ethical concerns with genome editing is the risk of off-target effects, where unintended modifications occur at genomic loci other than the intended target. Off-target effects can lead to unpredictable consequences, including genetic mutations, cellular dysfunction, and unintended phenotypic changes (Komor et al., 2017). Such unintended alterations raise questions about the safety and reliability of genome editing technologies, particularly in the context of human gene therapy and environmental release of edited organisms (Bolukbasi et al., 2016).

Germline Editing and Inheritable Genetic Modifications: The prospect of germline editing, where heritable genetic modifications are introduced into the germline cells of organisms, raises profound ethical questions. While germline editing offers the potential to prevent hereditary diseases and enhance human health, it also raises concerns about the long-term consequences and ethical implications of altering the genetic makeup of future generations (Lanphier et al., 2015). The ethical debate surrounding germline editing encompasses issues of consent, equity, and the unknown risks to future generations (Ishii, 2017).

Dual-Use Concerns and Biosecurity Risks: Genome editing technologies have dual-use potential, meaning they can be used for both beneficial and harmful purposes. While genome editing holds promise for therapeutic and agricultural applications, it also raises biosecurity risks, including the deliberate creation of bioweapons or the unintentional release of genetically modified organisms into the environment (DiEuliis et al., 2019). Addressing these dual-use concerns requires robust governance frameworks, international collaboration, and responsible research practices to mitigate potential risks (National

Academies of Sciences, Engineering, and Medicine, 2018).

Ethical Frameworks and Guiding Principles

To navigate the ethical complexities of genome editing, various ethical frameworks and guiding principles have been proposed. These frameworks aim to balance the potential benefits of genome editing with its ethical, social, and environmental implications. Key principles include:

Respect for Autonomy and Informed Consent: Respect for autonomy entails ensuring that individuals have the right to make informed decisions about their genetic information and participation in genome editing interventions (Savulescu et al., 2015). Informed consent mechanisms play a crucial role in upholding autonomy, requiring transparent communication of risks, benefits, and uncertainties associated with genome editing interventions.

Beneficence and Non-Maleficence: The principles of beneficence and non-maleficence emphasize the obligation to maximize the benefits of genome editing while minimizing harm (Hurlbut, 2015). This entails conducting rigorous risk assessments, prioritizing safety in research and clinical applications, and ensuring equitable access to genome editing technologies.

Justice and Equity: Justice and equity considerations underscore the importance of fair distribution of the benefits and burdens of genome editing technologies across diverse populations (Parens et al., 2016). This includes addressing disparities in access to healthcare, genetic therapies, and agricultural innovations, as well as safeguarding against unintended social, economic, and environmental consequences of genome editing interventions.

9.2 Regulatory Landscape for Non-Model Organisms

In recent years, the emergence of CRISPR/Cas technology has revolutionized genome editing, offering unprecedented precision and efficiency in modifying genetic material. However, alongside the scientific advancements, concerns about the ethical and regulatory implications of genome editing, particularly in non-model organisms, have come to the forefront. Understanding the regulatory landscape governing the use of CRISPR/Cas technology in non-model organisms is crucial for ensuring responsible research practices and compliance with legal requirements.

Regulatory Framework for Genome Editing

The regulatory framework for genome editing in non-model organisms varies widely across different countries and regions. In the United States, the primary regulatory agencies overseeing the use of genetically modified organisms (GMOs) include the Environmental Protection Agency (EPA), the Food and Drug Administration (FDA), and the Department of Agriculture (USDA). The regulatory process typically involves assessing the environmental and human health risks associated with genetically modified organisms and evaluating the potential benefits of the proposed modifications (Doudna & Charpentier, 2014).

Similarly, in the European Union (EU), the regulatory framework for genome editing is governed by the European Commission and the European Food Safety Authority (EFSA). The EU follows the precautionary principle, which requires a thorough risk assessment before approving the release of genetically modified organisms into the environment or the market (Wolt et al., 2016). Additionally, individual member states may have their own regulations governing the use of genetically modified organisms in research and agriculture.

Regulatory Considerations for Non-Model Organisms

Non-model organisms present unique regulatory challenges due to their diverse genetic backgrounds and ecological roles. Unlike model organisms such as mice or fruit flies, which have well-established genetic resources and research protocols, non-model organisms may lack comprehensive genomic information and may have limited research infrastructure (Carroll et al., 2016). Consequently, regulatory agencies may face difficulties in assessing the potential risks and benefits of genome editing in non-model organisms.

One of the key considerations in the regulatory evaluation of genome-edited non-model organisms is the potential for unintended ecological consequences. Non-model organisms often play critical roles in ecosystems, and genetic modifications could have unforeseen impacts on biodiversity, ecosystem dynamics, and ecosystem services (Waltz, 2016). For example, gene drive systems, which spread genetic modifications rapidly through wild populations, raise concerns about their potential to disrupt natural ecosystems and unintentionally harm non-target species (Esvelt et al., 2014).

Furthermore, the heritability and persistence of genome modifications in non-model organisms pose regulatory challenges. Unlike traditional genetic modification techniques, such as transgenesis, which typically involve the introduction of foreign DNA into the genome, CRISPR/Cas editing can induce precise modifications without the incorporation of exogenous DNA (Doudna & Charpentier, 2014). As a result, edited traits may be inherited by subsequent generations, raising questions about the long-term consequences of genome editing on population dynamics and evolutionary processes (Waltz, 2016).

Regulatory Responses to Genome Editing

In response to the growing use of genome editing technologies, regulatory agencies have begun to adapt their policies to address the unique challenges posed by these techniques. For example, the USDA recently announced updated regulations for genetically modified crops, which include provisions for the regulation of genome-edited organisms (Waltz, 2016). These regulations focus on the characteristics of the resulting organism rather than the process used to create it, distinguishing between organisms with and without foreign DNA insertions.

Similarly, the European Court of Justice ruled that organisms resulting from genome editing techniques should be subject to the same regulatory requirements as conventionally genetically modified organisms (Wolt et al., 2016). This decision reflects the EU's precautionary approach to risk assessment and emphasizes the importance of transparency and public engagement in the regulatory process.

However, the classification of genome-edited organisms under existing regulatory frameworks remains a topic of debate and uncertainty. Some stakeholders argue that organisms produced through precise genome editing techniques should be subject to less stringent regulations than those generated through traditional transgenic methods (Waltz, 2016). They argue that the targeted nature of CRISPR/Cas editing reduces the likelihood of unintended effects and therefore warrants a more flexible regulatory approach.

9.3 Guidelines and Best Practices

Alongside CRISPR/CAS advancements in genome editing come ethical and regulatory considerations, especially concerning the editing of non-model organisms. To ensure responsible use and minimize potential risks, various guidelines and best practices have been developed. In this section, we delve into the key guidelines and best practices for genome editing in non-model organisms, supported by evidence and data.

International Guidelines and Frameworks

The ethical and regulatory landscape surrounding genome editing is dynamic and varies across countries and regions. Nevertheless, several international organizations and scientific bodies have formulated overarching principles and guidelines.

The *International Commission on the Rules for the Approval of Genetically Engineered Organisms (ICRAOGEO)* provides a comprehensive framework for evaluating the risks and benefits of genetically engineered organisms, including those created through genome editing. Their guidelines emphasize the importance of risk assessment, transparency, and public engagement in decision-making processes (ICRAOGEO, 2020).

National Regulatory Agencies

In many countries, national regulatory agencies oversee the approval and regulation of genetically modified organisms (GMOs), including those generated using genome editing techniques. These agencies often develop guidelines and best practices tailored to their respective jurisdictions.

For example, the *United States Food and Drug Administration (FDA)* regulates genetically engineered organisms under the Coordinated Framework for Regulation of Biotechnology, which encompasses various federal agencies, including the FDA, the Environmental Protection Agency (EPA), and the Department of Agriculture (USDA) (FDA, 2020). The FDA evaluates the safety and environmental impact of genetically engineered organisms on a case-by-case basis, considering factors such as the intended use, characteristics of the organism, and potential ecological consequences.

Best Practices for Non-Model Organism Editing

Beyond regulatory oversight, there are several best practices that researchers should adhere to when conducting genome editing experiments in non-model organisms.

Risk Assessment and Mitigation: Before initiating genome editing experiments, researchers should conduct thorough risk assessments to identify potential hazards and mitigate associated risks. This includes assessing the likelihood of off-target effects, unintended ecological consequences, and impacts on biodiversity. Strategies for risk mitigation may include selecting target sites with minimal off-target potential, conducting rigorous environmental impact assessments, and implementing containment measures to prevent the spread of genetically modified organisms.

Transparency and Public Engagement: Transparency and public engagement are essential components of responsible genome editing research. Researchers should communicate openly about the goals, methods, and potential risks of their experiments with stakeholders, including the public, policymakers, and relevant regulatory agencies. Engaging with stakeholders early in the research process can help identify concerns, build trust, and foster informed decision-making.

Ethical Considerations: Ethical considerations should guide all stages of genome editing research, from experimental design to data interpretation and dissemination. Researchers should prioritize the welfare of non-model organisms, minimize harm, and respect principles of non-maleficence, beneficence, and justice. This may involve obtaining informed consent from stakeholders, considering the potential impacts on biodiversity and ecosystems, and upholding principles of scientific integrity and responsible conduct.

Data Sharing and Reproducibility: To promote transparency and reproducibility, researchers should share their data, methodologies, and findings openly with the scientific community. This includes depositing sequences in public databases, publishing results in peer-reviewed journals, and making protocols and reagents readily available to other researchers. By facilitating access to resources and promoting collaboration, data sharing can accelerate scientific progress and enhance the reliability of genome editing research.

Continuous Monitoring and Evaluation: Genome editing technologies are rapidly evolving, and our understanding of their risks and limitations continues to expand. Therefore, it is essential to continuously monitor and evaluate the safety and efficacy of genome editing techniques in non-model organisms. This may involve conducting long-term environmental monitoring studies, reassessing risk assessments in light of new evidence, and updating guidelines and best practices accordingly.

9.4 Public Perception and Engagement

Public perception and engagement play a crucial role in shaping the ethical and regulatory landscape surrounding genome editing in non-model organisms. Understanding public attitudes, concerns, and expectations is essential for policymakers, scientists, and stakeholders to develop responsible guidelines and regulations that align with societal values and preferences. In this section, we delve into the current state of public perception, the factors influencing it, and strategies for effective public engagement in the context of CRISPR/Cas editing in non-model organisms.

Public Perception of Genome Editing

Public perception of genome editing, including CRISPR/Cas technology, varies across different demographic groups, regions, and cultural contexts. Numerous surveys and studies have been conducted to gauge public attitudes towards genome editing in various organisms, shedding light on key concerns, perceptions, and acceptance levels.

A study by Xiang et al. (2020) surveyed public attitudes towards genome editing in non-model organisms in the United States, revealing a generally positive perception of the technology's potential benefits in agriculture and conservation efforts. However, concerns regarding environmental impacts, biodiversity loss, and potential unforeseen consequences were also prevalent among respondents.

Similarly, a global survey conducted by Scheufele et al. (2019) found that while the public generally supports the use of genome editing to address human health issues, such as curing genetic diseases, attitudes towards editing non-human organisms, including crops and animals, were more nuanced. Concerns related to environmental risks, unintended consequences, and ethical implications were cited as significant factors influencing public acceptance.

Factors Influencing Public Perception

Several factors influence public perception of genome editing in non-model organisms, including scientific literacy, trust in regulatory bodies and institutions, media portrayal, cultural and religious beliefs, and perceived risks and benefits.

Scientific literacy and understanding of genetic concepts significantly impact individuals' perceptions of genome editing. Studies have shown that individuals with higher levels of scientific knowledge tend to have more positive attitudes towards genetic technologies (Allum et al., 2017).

Trust in regulatory bodies and scientific institutions is another crucial factor shaping public perception. Transparency, accountability, and effective communication from scientists and policymakers are essential for building and maintaining public trust in genome editing research and applications (Hart Research Associates, 2018).

Media coverage and portrayal of genome editing technologies can influence public attitudes and perceptions. Sensationalized or inaccurate reporting may contribute to misconceptions, fear, and public distrust. Effective science communication efforts aimed at providing accurate information and addressing misconceptions are therefore paramount (Brossard & Lewenstein, 2010).

Cultural and religious beliefs also play a significant role in shaping public attitudes towards genome editing. Ethical considerations, such as respect for nature and concerns about playing "God," may influence individuals' views on the ethical permissibility of modifying non-human organisms (Fujimura & Rajagopalan, 2011).

Strategies for Effective Public Engagement

Effective public engagement is essential for fostering informed dialogue, addressing concerns, and building public trust and support for genome editing research and applications in non-model organisms. Scientists, policymakers, and stakeholders can employ various strategies to engage the public in meaningful discussions and decision-making processes.

Education and Outreach: Investing in public education and outreach initiatives to increase scientific literacy and understanding of genome editing technologies can help demystify complex scientific concepts and address misconceptions (National Academies of Sciences, Engineering, and Medicine, 2017).

Transparency and Openness: Ensuring transparency and openness in research and decision-making processes, including disclosing potential risks and uncertainties associated with genome editing, can enhance public trust and confidence in scientific endeavours (Pew Research Centre, 2018).

Public Consultation and Deliberation: Engaging the public in participatory decision-making processes through methods such as citizen juries, deliberative polling, and public forums can provide valuable insights, identify concerns, and foster dialogue between scientists, policymakers, and the public (Burgess et al., 2020).

Ethical and Social Impact Assessments: Conducting ethical and social impact assessments of genome editing research and applications, involving diverse stakeholders, can help identify and address ethical, cultural, and societal implications, guiding responsible innovation and policy development (Nuffield Council on Bioethics, 2016).

Cultural Sensitivity and Inclusivity: Recognizing and respecting cultural diversity and values when communicating about genome editing technologies is essential for engaging diverse communities and addressing cultural concerns and perspectives (Doudna, 2020).

References

Adamson, B., Norman, T. M., Jost, M., Cho, M. Y., Nuñez, J. K., Chen, Y., ... & Weissman, J. S. (2016). A multiplexed single-cell CRISPR screening platform enables systematic dissection of the unfolded protein response. Cell, 167(7), 1867-1882.

Adli, M. (2018). The CRISPR tool kit for genome editing and beyond. *Nature communications*, 9(1), 1911.

Akinc, A., Maier, M. A., Manoharan, M., Fitzgerald, K., Jayaraman, M., Barros, S., ... & Koteliansky, V. (2019). The Onpattro story and the clinical translation of nanomedicines containing nucleic acid-based drugs. Nature nanotechnology, 14(12), 1084-1087.

Allum, N., Sturgis, P., Tabourazi, D., & Brunton-Smith, I. (2017). Science knowledge and attitudes across cultures: A meta-analysis. Public Understanding of Science, 26(6), 742–759.

Altpeter, F., Springer, N. M., Bartley, L. E., Blechl, A. E., Brutnell, T. P., Citovsky, V., ... & Datta, S. K. (2016). Advancing crop transformation in the era of genome editing. The Plant Cell, 28(7), 1510-1520.

Altschul, S. F., Gish, W., Miller, W., Myers, E. W., & Lipman, D. J. (1990). Basic local alignment search tool. Journal of Molecular Biology, 215(3), 403–410.

Anzalone, A. V., Randolph, P. B., Davis, J. R., Sousa, A. A., Koblan, L. W., Levy, J. M., ... & Liu, D. R. (2019). Search-and-replace genome editing without double-strand breaks or donor DNA. Nature, 576(7785), 149-157.

Ashburner, M., Ball, C. A., Blake, J. A., Botstein, D., Butler, H., Cherry, J. M., ... & Sherlock, G. (2000). Gene ontology: tool for the unification of biology. Nature Genetics, 25(1), 25–29.

Bae, S., Park, J., & Kim, J. (2014). Cas-OFFinder: a fast and versatile algorithm that searches for potential off-target sites of Cas9 RNA-guided endonucleases. Bioinformatics, 30(10), 1473–1475.

Bae, S., Park, J., & Kim, J.-S. (2014). Cas-OFFinder: A fast and versatile algorithm that searches for potential off-target sites of Cas9 RNA-guided endonucleases. Bioinformatics, 30(10), 1473–1475.

Bae, S., Park, J., & Kim, J.-S. (2014). Cas-OFFinder: A fast and versatile algorithm that searches for potential off-target sites of Cas9 RNA-guided endonucleases. Bioinformatics, 30(10), 1473–1475.

Barrangou, R., & Doudna, J. A. (2016). Applications of CRISPR technologies in research and beyond. Nature Biotechnology, 34(9), 933–941.

Barrangou, R., Fremaux, C., Deveau, H., Richards, M., Boyaval, P., Moineau, S., ... & Horvath, P. (2007). CRISPR provides acquired resistance against viruses in prokaryotes. Science, 315(5819), 1709-1712.

Benchling. CRISPR/Cas9 Genome Editing (n.d.). Accessed February 28, 2024. https://www.benchling.com/crispr/

Bolukbasi, M. F., Gupta, A., & Wolfe, S. A. (2016). Creating and evaluating accurate CRISPR-Cas9 scalpels for genomic surgery. Nature Methods, 13(1), 41–50.

Borges, A. L., Davidson, A. R., & Bondy-Denomy, J. (2019). The discovery, mechanisms, and evolutionary impact of anti-CRISPRs. Annual Review of Virology, 6, 37-59.

Brossard, D., & Lewenstein, B. V. (2010). A critical appraisal of models of public understanding of science: Using practice to inform theory. In L. Kahlor & P. Stout (Eds.), Communicating Science: New Agendas in Communication (pp. 11–39). Routledge.

Burgess, M. M., Conti, N., Dudo, A., Ishiyama, J., & Leask, J. (2020). Engaging with science in social and digital spaces: A roadmap for science communication research. Journal of Science Communication, 19(01), A01_1–A01_18.

Cameron, P., & Vyas, R. (2020). CRISPR-Cas: An effective tool for genome engineering of non-model organisms. Journal of Experimental Biology, 223(Pt Suppl 1), jeb208700.

Chapman, J. R., Taylor, M. R., & Boulton, S. J. (2012). Playing the end game: DNA double-strand break repair pathway choice. Molecular Cell, 47(4), 497–510. doi: 10.1016/j.molcel.2012.07.029

Chen, B., Gilbert, L. A., Cimini, B. A., Schnitzbauer, J., Zhang, W., Li, G. W., Park, J., Blackburn, E. H.,

Weissman, J. S., Qi, L. S., et al. (2013). Dynamic imaging of genomic loci in living human cells by an optimized CRISPR/Cas system. Cell, 155(7), 1479–1491.

Chen, K., Wang, Y., Zhang, R., Zhang, H., Gao, C., & Li, J. (2020). CRISPR/Cas Genome Editing and Precision Plant Breeding in Agriculture. Annual Review of Plant Biology, 71, 683–707. [APA format]

Chen, S., Sanjana, N. E., Zheng, K., Shalem, O., Lee, K., Shi, X., ... & Sharp, P. A. (2015). Genome-wide CRISPR screen in a mouse model of tumor growth and metastasis. Cell, 160(6), 1246-1260.

Chen, S., Sanjana, N. E., Zheng, K., Shalem, O., Lee, K., Shi, X., ... & Doench, J. G. (2018). Genome-wide CRISPR screen in a mouse model of tumor growth and metastasis. Cell, 160(6), 1246-1260.

Chen, S., Sanjana, N. E., Zheng, K., Shalem, O., Lee, K., Shi, X., Scott, D. A., Song, J., Pan, J. Q., Weissleder, R., Lee, H., Zhang, F., & Sharp, P. A. (2020). Genome-wide CRISPR screen in a mouse model of tumor growth and metastasis.

Cell, 160(6), 1246–1260. doi:10.1016/j.cell.2015.02.038

Chen, S., Wang, Q., Lin, L., Lin, Y., Yang, S., et al. (2021). CRISPR/Cas9-mediated gene editing in zebrafish using a multi-omics approach. *Genomics*, 113(3), 1396-1403.

Chen, S., Yang, L., Hu, Q., & Hu, M. (2019). Nutrient limitation and stationary phase enhance genetic transformation of soil bacteria by electroporation. Applied and Environmental Microbiology, 85(4), e02513-18.

Chen, X., Gonçalves, M. A., & Mermoud, J. E. (2016). Efficient, footprint-free human iPSC genome editing by consolidation of Cas9/CRISPR and piggyBac technologies. Nature Protocols, 11(12), 2371-2388.

Cheng, A. W., Wang, H., Yang, H., Shi, L., Katz, Y., Theunissen, T. W., ... & Jaenisch, R. (2013). Multiplexed activation of endogenous genes by CRISPR-on, an RNA-guided transcriptional activator system. Cell research, 23(10), 1163-1171.

Chew, W. L., Tabebordbar, M., Cheng, J. K., Mali, P., Wu, E. Y., Ng, A. H., ... & Church, G. M. (2016).

A multifunctional AAV–CRISPR–Cas9 and its host response. Nature Methods, 13(10), 868-874.

Choi, H. M. T., Beck, V. A., & Pierce, N. A. (2021). Next-generation in vivo modeling of CRISPR/Cas-mediated gene editing in Drosophila melanogaster using integrative barcoded sgRNA libraries. ACS Synthetic Biology, 10(2), 271-280.

Clement, K., Rees, H., Canver, M. C., Gehrke, J. M., Farouni, R., et al. (2019). CRISPResso2 provides accurate and rapid genome editing sequence analysis. *Nature Biotechnology*, 37(3), 224-226.

Clement, K., Rees, H., Canver, M. C., Gehrke, J. M., Farouni, R., Hsu, J. Y., ... Bauer, D. E. (2019). CRISPResso2 provides accurate and rapid genome editing sequence analysis. Nature Biotechnology, 37(3), 224–226.

Cong, L., Ran, F. A., Cox, D., Lin, S., Barretto, R., Habib, N., ... & Zhang, F. (2013). Multiplex genome engineering using CRISPR/Cas systems. Science, 339(6121), 819-823.

Cong, L., Ran, F. A., Cox, D., Lin, S., Barretto, R., Habib, N., ... & Zhang, F. (2013). Multiplex genome engineering using CRISPR/Cas systems. Science, 339(6121), 819–823.

Dahlman, J. E., Abudayyeh, O. O., Joung, J., Gootenberg, J. S., Zhang, F., & Konermann, S. (2015). Orthogonal gene knockout and activation with a catalytically active Cas9 nuclease. Nature Biotechnology, 33(11), 1159–1161.

Didiot, M. C., Hall, L. M., Coles, A. H., Haraszti, R. A., Godinho, B. M., Chase, K., ... & Khvorova, A. (2016). Exosome-mediated delivery of hydrophobically modified siRNA for Huntingtin mRNA silencing. Molecular Therapy, 24(10), 1836-1847.

DiEuliis, D., Johnson, K. J., & Gronvall, G. K. (2019). Genomic biosecurity: Reconceptualizing dual-use. Frontiers in Bioengineering and Biotechnology, 7, 221.

Doench, J. G., Fusi, N., Sullender, M., Hegde, M., Vaimberg, E. W., Donovan, K. F., & Smith, I. (2016). Optimized sgRNA design to maximize activity and minimize off-target effects of

CRISPR-Cas9. *Nature biotechnology, 34*(2), 184–191.

Doench, J. G., Hartenian, E., Graham, D. B., Tothova, Z., Hegde, M., Smith, I., ... & Root, D. E. (2014). Rational design of highly active sgRNAs for CRISPR-Cas9–mediated gene inactivation. *Nature biotechnology, 32*(12), 1262-1267.

Doudna, J. A. (2020). CRISPR-Cas9 basics: Molecular biology and principles of genome editing. John Wiley & Sons.

Doudna, J. A., & Charpentier, E. (2014). Genome editing. The new frontier of genome engineering with CRISPR-Cas9. Science, 346(6213), 1258096.

Doudna, J. A., & Charpentier, E. (2014). The new frontier of genome engineering with CRISPR-Cas9. Science, 346(6213), 1258096.

Doudna, J. A., & Charpentier, E. (2014). The new frontier of genome engineering with CRISPR-Cas9. Science, 346(6213), 1258096. Esvelt, K. M., Smidler, A. L., Catteruccia, F., & Church, G. M. (2014). Concerning RNA-guided gene drives for the alteration of wild populations. Elife, 3, e03401. Carroll, D., Charo, R. A., & Copenhaver, G. P. (2016). Genome editing. Science, 351(6271), 141. Waltz, E. (2016). Gene-edited CRISPR mushroom escapes US regulation. Nature, 532(7599), 293. Wolt, J. D., Wang, K., Yang, B., & Spalding, M. H. (2016). Regulatory aspects of genome-edited crops in the United States. Plant Biotechnology Journal, 14(2), 510-518.

ENCODE Project Consortium. (2012). An integrated encyclopedia of DNA elements in the human genome. Nature, 489(7414), 57–74.

Fauser, F., Schiml, S., & Puchta, H. (2012). Both CRISPR/Cas-based nucleases and nickases can be used efficiently for genome engineering in Arabidopsis thaliana. The Plant Journal, 79(2), 348–359.

Feng, Z., Zhang, B., Ding, W., Liu, X., Yang, D. L., Wei, P., ... & Zhu, J. K. (2013). Efficient genome editing in plants using a CRISPR/Cas system. Cell research, 23(10), 1229-1232.

Fisher, J., McNulty, M., & Cowman, P. F. (2021). Environmental adaptation and habitat-specific selection in a diverse marine community. Ecology Letters, 24(1), 189-196.

Frock, R. L., Hu, J., Meyers, R. M., Ho, Y. J., Kii, E., & Alt, F. W. (2015). Genome-wide detection of DNA double-stranded breaks induced by engineered nucleases. Nature Biotechnology, 33(2), 179–186.

Fu, Y., Foden, J. A., Khayter, C., Maeder, M. L., Reyon, D., Joung, J. K., & Sander, J. D. (2014). High-frequency off-target mutagenesis induced by CRISPR-Cas nucleases in human cells. Nature biotechnology, 31(9), 822–826.

Fu, Y., Sander, J. D., Reyon, D., Cascio, V. M., & Joung, J. K. (2014). Improving CRISPR-Cas nuclease specificity using truncated guide RNAs. Nature Biotechnology, 32(3), 279–284.

Fujimura, J. H., & Rajagopalan, R. (2011). Different differences: The use of 'genetic ancestry' versus race in biomedical human genetic research. Social Studies of Science, 41(1), 5–30.

Gagnon, J. A., Valen, E., Thyme, S. B., Huang, P., Ahkmetova, L., Pauli, A., ... & Schier, A. F. (2014). Efficient mutagenesis by Cas9 protein-mediated oligonucleotide insertion and large-scale assessment of single-guide RNAs. PloS one, 9(5), e98186.

Gaj, T., Gersbach, C. A., & Barbas, C. F. (2013). ZFN, TALEN, and CRISPR/Cas-based methods for genome engineering. Trends in biotechnology, 31(7), 397-405.

Gaj, T., Gersbach, C. A., & Barbas, C. F. (2016). ZFN, TALEN, and CRISPR/Cas-based methods for genome engineering. Trends in Biotechnology, 31(7), 397–405.

Garcia, A., Smith, J., & Patel, M. (2023). High-throughput screening of guide RNAs for CRISPR/Cas editing in the sea urchin Strongylocentrotus purpuratus. Marine Genomics, 57, 100880.

Gasperini, M., Hill, A. J., McFaline-Figueroa, J. L., Martin, B., Kim, S., Zhang, M. D., ... & Shendure, J. (2019). A genome-wide framework for mapping gene regulation via cellular genetic screens. *Cell, 176*(1), 377-390.

Gomez, J. A., Wapinski, O. L., Yang, Y. W., Bureau, J. F., & Gifford, D. K. (2021). The Neanderthal Genomic Landscape and Its Impact on Gene

Expression in Human Populations. Cell, 184(10), 2641-2653.

Gori, J. L., Hsu, P. D., Maeder, M. L., Shen, S., Welstead, G. G., Bumcrot, D., & Delivery of genome editing tools via lentiviral vectors. (2015). Delivery of genome editing tools via lentiviral vectors. Bioengineering & Translational Medicine, 1(2), 123-131.

Haeussler M, Schönig K, Eckert H, et al. Evaluation of off-target and on-target scoring algorithms and integration into the guide RNA selection tool CRISPOR. Genome Biol. 2016;17(1):148. doi:10.1186/s13059-016-1012-2

Haeussler, M., Schönig, K., Eckert, H., Eschstruth, A., Mianné, J., Renaud, J. B., ... & Schneider-Maunoury, S. (2016). Evaluation of off-target and on-target scoring algorithms and integration into the guide RNA selection tool CRISPOR. Genome Biology, 17(1), 148.

Haeussler, M., Schönig, K., Eckert, H., Eschstruth, A., Mianné, J., Renaud, J.-B., ... & Schneider-Maunoury, S. (2016). Evaluation of off-target and on-target scoring algorithms and integration into the guide RNA selection tool CRISPOR. Genome Biology, 17(1), 148.

Hart Research Associates. (2018). Public attitudes toward gene editing. Retrieved from https://www.asbmb.org/docs/default-source/about-us/policy-statements/asbmb-crispr-survey-data-and-memo---december-2018.pdf

Heigwer, F., Kerr, G., & Boutros, M. (2014). E-CRISP: fast CRISPR target site identification. Nature Methods, 11(2), 122-123.

Hruscha, A., Krawitz, P., Rechenberg, A., Heinrich, V., Hecht, J., Haass, C., & Schmid, B. (2013). Efficient CRISPR/Cas9 genome editing with low off-target effects in zebrafish. Development, 140(24), 4982-4987.

Hsu PD, Lander ES, Zhang F. Development and applications of CRISPR-Cas9 for genome engineering. Cell. 2014;157(6):1262-1278. doi:10.1016/j.cell.2014.05.010

Hsu, P. D., Lander, E. S., & Zhang, F. (2014). Development and applications of CRISPR-Cas9 for genome engineering. Cell, 157(6), 1262-1278.

Hsu, P. D., Scott, D. A., Weinstein, J. A., Ran, F. A., Konermann, S., Agarwala, V., & Zhang, F. (2013). DNA targeting specificity of RNA-guided Cas9 nucleases. Nature biotechnology, 31(9), 827–832.

Hsu, P. D., Scott, D. A., Weinstein, J. A., Ran, F. A., Konermann, S., Agarwala, V., ... & Zhang, F. (2013). DNA targeting specificity of RNA-guided Cas9 nucleases. *Nature biotechnology, 31*(9), 827-832.

Hsu, P. D., Scott, D. A., Weinstein, J. A., Ran, F. A., Konermann, S., Agarwala, V., ... & Zhang, F. (2013). DNA targeting specificity of RNA-guided Cas9 nucleases. Nature Biotechnology, 31(9), 827–832.

Hsu, P. D., Scott, D. A., Weinstein, J. A., Ran, F. A., Konermann, S., Agarwala, V., ... & Zhang, F. (2013). DNA targeting specificity of RNA-guided Cas9 nucleases. Nature Biotechnology, 31(9), 827–832.

Hurlbut, J. B. (2015). Human dignity and the future of man. In A. J. Caplan, G. McLean, & B. P. Baum (Eds.), Ethics and the acquisition of organs (pp. 19–45). Springer.

Ishii, T. (2017). Germline genome-editing research and its socioethical implications. Trends in Molecular Medicine, 23(10), 835–848.

Ishii, T., & Araki, M. (2016). A future scenario of the global regulatory landscape regarding genome-edited crops. GM Crops & Food, 7(1), 1-10.

Jackson, S. P., & Bartek, J. (2009). The DNA-damage response in human biology and disease. Nature, 461(7267), 1071–1078.

Javed, M. R., Poschmann, G., Nadeem, M. S., Roschitzki, B., Lercher, M. J., et al. (2020). CRISPR-FOCUS: A web server for designing focused CRISPR screening experiments. *PLoS Computational Biology*, 16(9), e1008200.

Jiang, W., Bikard, D., Cox, D., Zhang, F., & Marraffini, L. A. (2013). RNA-guided editing of bacterial genomes using CRISPR-Cas systems. Nature biotechnology, 31(3), 233-239.

Jinek, M., Chylinski, K., Fonfara, I., Hauer, M., Doudna, J. A., & Charpentier, E. (2012). A programmable dual-RNA-guided DNA endonuclease in adaptive bacterial immunity. Science, 337(6096), 816–821.

Jinek, M., Chylinski, K., Fonfara, I., Hauer, M., Doudna, J. A., & Charpentier, E. (2012). A programmable dual-RNA-guided DNA endonuclease in adaptive bacterial immunity. Science, 337(6096), 816-821.

Jinek, M., Chylinski, K., Fonfara, I., Hauer, M., Doudna, J. A., & Charpentier, E. (2012). A programmable dual-RNA–guided DNA endonuclease in adaptive bacterial immunity. Science, 337(6096), 816–821.

Johnson, R., Lindblad-Toh, K., & Liu, H. (2020). High-throughput CRISPR screening in zebrafish identifies gRNAs targeting non-coding elements for highly efficient genome editing. BMC Genomics, 21(1), 1-12.

Jones, F. C., Grabherr, M. G., Chan, Y. F., Russell, P., Mauceli, E., Johnson, J., ... & Kingsley, D. M. (2012). The genomic basis of adaptive evolution in threespine sticklebacks. Nature, 484(7392), 55-61.

Jones, J. R., Mir, K. K., Hyeong, C. C., Wu, L., & David, G. A. (2020). Dynamic transcriptional and epigenetic regulation of human epidermal keratinocyte differentiation. Cell Stem Cell, 22(1), 10-24.

Jones, M. R., Blanca, J., Morikawa, Y., Cosio, E. G., & Chang, S. M. (2019). Whole-genome sequence and assembly of *Arabidopsis lyrata* subsp. *kamchatica*, a model for the study of polyploidy. Genome Biology, 20(1), 29. [APA format]

Jones, M., & Patel, S. (2024). Expanding the CRISPR toolkit: Applications of base editing and prime editing in non-model organisms. Trends in Biotechnology.
doi:10.1016/j.tibtech.2023.12.006

Kanehisa, M., & Goto, S. (2000). KEGG: Kyoto encyclopedia of genes and genomes. Nucleic Acids Research, 28(1), 27–30.

Kent, W. J., Sugnet, C. W., Furey, T. S., Roskin, K. M., Pringle, T. H., Zahler, A. M., & Haussler, D. (2002). The human genome browser at UCSC. Genome Research, 12(6), 996–1006.

Kim, D., Bae, S., Park, J., Kim, E., Kim, S., Yu, H. R., & Hwang, J. (2018). Digenome-seq: genome-wide profiling of CRISPR-Cas9 off-target effects in human cells. Nature Methods, 15(4), 275–278.

Kim, D., Bae, S., Park, J., Kim, E., Kim, S., Yu, H. R., ... & Kim, J. S. (2015). Digenome-seq: genome-wide profiling of CRISPR-Cas9 off-target effects in human cells. Nature Methods, 12(3), 237-243.

Kleinstiver, B. P., Prew, M. S., Tsai, S. Q., Topkar, V. V., Nguyen, N. T., Zheng, Z., ... Joung, J. K. (2019). Engineered CRISPR-Cas12a variants with increased activities and improved targeting ranges for gene, epigenetic and base editing. Nature Biotechnology, 37(3), 276–282. doi: 10.1038/s41587-018-0011-0

Komor, A. C., Kim, Y. B., Packer, M. S., Zuris, J. A., & Liu, D. R. (2016). Programmable editing of a target base in genomic DNA without double-stranded DNA cleavage. Nature, 533(7603), 420-424.

Komor, A. C., Kim, Y. B., Packer, M. S., Zuris, J. A., & Liu, D. R. (2017). Programmable editing of a target base in genomic DNA without double-stranded DNA cleavage. Nature, 533(7603), 420–424.

Konermann, S., Brigham, M. D., Trevino, A. E., Hsu, P. D., Heidenreich, M., Cong, L., & Zhang, F. (2015). Optical control of mammalian endogenous transcription and epigenetic states. Nature, 500(7463), 472–476.

Kosicki M, Tomberg K, Bradley A. Repair of double-strand breaks induced by CRISPR-Cas9 leads to large deletions and complex rearrangements. Nat Biotechnol. 2018;36(8):765-771. doi:10.1038/nbt.4192

Labun, K., Montague, T. G., Gagnon, J. A., Thyme, S. B., & Valen, E. (2019). CHOPCHOP v3: expanding the CRISPR web toolbox beyond genome editing. Nucleic Acids Research, 47(W1), W171-W174.

Labun, K., Montague, T. G., Gagnon, J. A., Thyme, S. B., & Valen, E. (2016). CHOPCHOP v2: A web tool for the next generation of CRISPR genome engineering. Nucleic Acids Research, 44(W1), W272–W276.

Langmead, B., Trapnell, C., Pop, M., & Salzberg, S. L. (2009). Ultrafast and memory-efficient alignment of short DNA sequences to the human genome. Genome Biology, 10(3), R25.

Lanphier, E., Urnov, F., Haecker, S. E., Werner, M., & Smolenski, J. (2015). Don't edit the human germ line. Nature, 519(7544), 410–411.

Lee, H., Hwang, W., Lee, Y., Park, J. Y., Park, J., Bae, J. Y., ... & Han, K. A. (2021). Optimization of CRISPR/Cas9 and its gRNA in Drosophila suzukii for genome editing. PLoS One, 16(4), e0250001.

Lei, Y., Lu, L., Liu, H. Y., Li, S., Xing, F., et al. (2014). CRISPR-P: A web tool for synthetic single-guide RNA design of CRISPR-system in plants. *Molecular Plant*, 7(9), 1494-1496.

Li, A., Wetzel, D., Li, L., & Den Hollander, J. A. I. (2018). High-throughput, CRISPR-based gene editing for Aspergillus Niger and the selection of editotypes for hyper-production of commercially relevant proteins. Biotechnology and Bioengineering, 115(7), 1568-1576.

Li, G., Liu, S., & Yang, H. (2019). High-throughput identification of small-molecule enhancers of CRISPR/Cas9-mediated homology-directed repair. Cell Reports, 26(4), 1098-1109.

Li, G., Zhang, X., Zhong, C., Mo, J., Quan, R., Yang, J., ... Wei, Y. (2019). Small Size of Donor DNA

Fragments Promotes HDR in the Cells of Plants. Frontiers in Plant Science, 10, 1676. doi: 10.3389/fpls.2019.01676

Li, H., Li, C., Li, T., Yao, W., Wang, Y., & Wang, X. (2020). Investigating the effects of pH on the efficiency of CRISPR/Cas9 gene editing in marine microalgae. Algal Research, 47, 101874.

Li, H., Li, J., Meng, X., Gan, L., Zhang, R., Gao, C., & Li, J. (2021). Precise modifications of both exogenous and endogenous genes in *Tribolium castaneum* through a simple and efficient CRISPR/Cas9-based method. Insect Science. Advance online publication.

Li, M., Zhao, L., Page-McCaw, P. S., Chen, W., Zhou, Y., Wang, H., ... & Wang, L. (2016). Targeted gene disruption by CRISPR/Cas9 system in the model fish medaka. Biological Open, 5(6), 858-863.

Li, X., Sandri-Goldin, R. M., & Zheng, Z. M. (2019). Herpes simplex virus: transcriptional regulation, DNA replication, and the latency-reactivation cycle. Cell Research, 29(1), 1-20.

Li, X., Zhu, C., Yeh, C. T., Wu, W., Takacs, E. M., Petsch, K. A., ... & Timmermans, M. C. (2020).

Genic and global functions for Paf1C in chromatin modification and gene expression in Drosophila melanogaster. PLOS Genetics, 16(3), e1008623.

Liang, X., Potter, J., Kumar, S., Zou, Y., Quintanilla, R., Sridharan, M., ... & DeKelver, R. C. (2015). Rapid and highly efficient mammalian cell engineering via Cas9 protein transfection. Journal of Biotechnology, 208, 44-53.

Liu, H., Wei, Z., Dominguez, A., Li, Y., Wang, X., & Qi, L. S. (2015). CRISPR-ERA: a comprehensive design tool for CRISPR-mediated gene editing, repression and activation. *Bioinformatics, 31*(22), 3676-3678.

Liu, J., Sun, Y., Roesler, A., Kariyawasam, T., Wang, G., & Zhang, W. (2020). CRISPR/Cas9 editing of endogenous banana streak virus in the B genome of Musa spp. overcomes a major challenge in banana breeding. Communications Biology, 3(1), 1–10.

Liu, S. J., Horlbeck, M. A., Cho, S. W., Birk, H. S., Malatesta, M., He, D., & Kampmann, M. (2017). CRISPRi-based genome-scale identification of functional long noncoding

RNA loci in human cells. Science, 355(6320), eaah7111.

Liu, Y., Gao, H., Marois, E., & Dong, S. (2018). CRISPR/Cas9-mediated genome editing in the crustacean Daphnia magna. *Scientific Reports*, 8(1), 1-9.

Liu, Y., Ye, N., & Huang, B. (2021). Temperature-dependent CRISPR/Cas9 mediated genome editing in the coral Acropora millepora. Marine Genomics, 56, 100780. doi:10.1016/j.margen.2020.100780

Liu, Y., Zhan, Y., Chen, Z., He, A., Li, J., Wu, H., ... & Li, L. (2019). Directing cellular information flow via CRISPR signal conductors. Nature Methods, 16(12), 1111-1114.

Maeder, M. L., Angstman, J. F., Richardson, M. E., Linder, S. J., Cascio, V. M., Tsai, S. Q., ... & Joung, J. K. (2013). Targeted DNA demethylation and activation of endogenous genes using programmable TALE-TET1 fusion proteins. Nature biotechnology, 31(12), 1137-1142.

Mali, P., Yang, L., Esvelt, K. M., Aach, J., Guell, M., DiCarlo, J. E., ... & Church, G. M. (2013). RNA-guided human genome engineering via Cas9. Science, 339(6121), 823-826.

Mali, P., Yang, L., Esvelt, K. M., Aach, J., Guell, M., DiCarlo, J. E., ... & Church, G. M. (2013). RNA-guided human genome engineering via Cas9. Science, 339(6121), 823-826.

Mali, P., Yang, L., Esvelt, K. M., Aach, J., Guell, M., DiCarlo, J. E., ... & Church, G. M. (2013). RNA-guided human genome engineering via Cas9. Science, 339(6121), 823–826.

Mao, Z., Bozzella, M., Seluanov, A., & Gorbunova, V. (2019). DNA repair by nonhomologous end joining and homologous recombination during cell cycle in human cells. Cell Cycle, 7(18), 2902–2906. doi: 10.1080/15384101.2018.1548529

Maruyama, T., Dougan, S. K., Truttmann, M. C., Bilate, A. M., Ingram, J. R., & Ploegh, H. L. (2015). Increasing the efficiency of precise genome editing with CRISPR-Cas9 by inhibition of nonhomologous end joining. Nature Biotechnology, 33(5), 538–542. doi: 10.1038/nbt.3190

Maruyama, T., Dougan, S. K., Truttmann, M. C., Bilate, A. M., Ingram, J. R., & Ploegh, H. L. (2015). Increasing the efficiency of precise genome editing with CRISPR-Cas9 by inhibition of nonhomologous end joining. Nature Biotechnology, 33(5), 538–542.

McCarty, D. M., Monahan, P. E., & Samulski, R. J. (2003). Self-complementary recombinant adeno-associated virus (scAAV) vectors promote efficient transduction independently of DNA synthesis. Gene Therapy, 10(26), 2112-2118.

Mingozzi, F., & High, K. A. (2013). Immune responses to AAV vectors: overcoming barriers to successful gene therapy. Blood, 122(1), 23-36.

Mojica, F. J., Díez-Villaseñor, C., García-Martínez, J., & Soria, E. (2005). Intervening sequences of regularly spaced prokaryotic repeats derive from foreign genetic elements. Journal of molecular evolution, 60(2), 174-182.

Montague, T. G., Cruz, J. M., Gagnon, J. A., Church, G. M., & Valen, E. (2014). CHOPCHOP: a CRISPR/Cas9 and TALEN web tool for genome editing. *Nucleic acids research*, 42(W1), W401-W407.

Moreno-Mateos, M. A., Vejnar, C. E., Beaudoin, J. D., Fernandez, J. P., Mis, E. K., Khokha, M. K., & Giraldez, A. J. (2015). CRISPRscan: designing highly efficient sgRNAs for CRISPR-Cas9 targeting in vivo. Nature Methods, 12(10), 982-988.

Moreno-Mateos, M. A., Vejnar, C. E., Beaudoin, J. D., Fernandez, J. P., Mis, E. K., Khokha, M. K., & Giraldez, A. J. (2015). CRISPRscan: designing highly efficient sgRNAs for CRISPR-Cas9 targeting in vivo. Nature Methods, 12(10), 982-988.

Moreno-Mateos, M. A., Vejnar, C. E., Beaudoin, J. D., Fernandez, J. P., Mis, E. K., Khokha, M. K., & Giraldez, A. J. (2015). CRISPRscan: Designing highly efficient sgRNAs for CRISPR-Cas9 targeting in vivo. Nature Methods, 12(10), 982–988.

Moreno-Mateos, M. A., Vejnar, C. E., Beaudoin, J. D., Fernandez, J. P., Mis, E. K., Khokha, M. K., & Giraldez, A. J. (2015). CRISPRscan: designing highly efficient sgRNAs for CRISPR-Cas9 targeting in vivo. Nature Methods, 12(10), 982–988.

Naito, Y., Hino, K., Bono, H., & Ui-Tei, K. (2015). CRISPRdirect: software for designing CRISPR/Cas guide RNA with reduced off-target sites. Bioinformatics, 31(7), 1120–1123.

Nakade, S., Tsubota, T., Sakane, Y., Kume, S., Sakamoto, N., Obara, M., ... Sakuma, T. (2014). Microhomology-mediated end-joining-dependent integration of donor DNA in cells and animals using TALENs and CRISPR/Cas9. Nature Communications, 5, 5560. doi: 10.1038/ncomms6560

National Academies of Sciences, Engineering, and Medicine. (2017). Human genome editing: Science, ethics, and governance. National Academies Press.

National Academies of Sciences, Engineering, and Medicine. (2018). Biodefense in the age of synthetic biology. The National Academies Press.

Nuffield Council on Bioethics. (2016). Genome editing: An ethical review. Retrieved from https://www.nuffieldbioethics.org/publications/genome-editing-an-ethical-review

Paix, A., Folkmann, A., Goldman, D. H., Kulaga, H., Grzelak, M. J., Rasoloson, D., Paidemarry, S., Green, R., Reed, R. R., Seydoux, G., & (2017). Precision genome editing using synthesis-dependent repair of Cas9-induced DNA breaks. Proceedings of the National Academy of Sciences, 114(50), E10745–E10754.

Paix, A., Folkmann, A., Rasoloson, D., & Seydoux, G. (2015). High efficiency, homology-directed genome editing in Caenorhabditis elegans using CRISPR-Cas9 ribonucleoprotein complexes. Genetics, 201(1), 47-54.

Parens, E., Johnston, J., & Moses, J. (2016). Ethical issues in germline gene editing. Science, 348(6237), 36–38.

Park, J., Kim, J. S., & Bae, S. (2017). Cas-Designer: A web-based tool for choice of CRISPR-Cas9 target sites. Bioinformatics, 33(24), 4015-4017.

Park, J., Lim, K., Kim, J. S., & Bae, S. (2017). Cas-analyzer: an online tool for assessing genome editing results using NGS data. Bioinformatics, 33(2), 286-288.

Peng, D., & Tarleton, R. (2015). EuPaGDT: a web tool tailored to design CRISPR guide RNAs for eukaryotic pathogens. *Microbial genomics, 1*(4), e000033.

Peng, X., Xiong, X., & Chen, L. (2018). Wang, X. Application of the CRISPR/Cas9 system to Chlamydomonas reinhardtii. Acta Oceanologica Sinica, 37(9), 23–29.

Pew Research Center. (2018). U.S. public wary of biomedical technologies to 'enhance' human abilities. Retrieved from https://www.pewresearch.org/science/2018/07/26/u-s-public-wary-of-biomedical-technologies-to-enhance-human-abilities/

Pineda, M., & Moghadam, H. K. (2019). Applications of CRISPR technology in marine biology and aquaculture. Fish and Shellfish Immunology, 86, 1056–1066.

Pinello, L., Canver, M. C., Hoban, M. D., Orkin, S. H., & Bauer, D. E. (2016). CRISPResso: A software for genome editing analysis. Nature Biotechnology, 34(7), 695–697.

Pinello, L., Canver, M. C., Hoban, M. D., Orkin, S. H., Kohn, D. B., Bauer, D. E., & Yuan, G. C. (2016).

Analyzing CRISPR genome-editing experiments with CRISPResso. Nature Biotechnology, 34(7), 695-697.

Port, F., Chen, H. M., Lee, T., & Bullock, S. L. (2014). Optimized CRISPR/Cas tools for efficient germline and somatic genome engineering in Drosophila. Proceedings of the National Academy of Sciences, 111(29), E2967-E2976.

Puchta, H., Fauser, F., & GeneViToP Consortium. (2021). Synthetic biology for directed crop improvement. Plant Journal, 105(4), 913-928.

Rahdar, M., McMahon, M. A., Prakash, T. P., Swayze, E. E., Bennett, C. F., & Cleveland, D. W. (2015). Synthetic CRISPR RNA-Cas9–guided genome editing in human cells. Proceedings of the National Academy of Sciences, 112(51), E7110–E7117.

Ran, F. A., Cong, L., Yan, W. X., Scott, D. A., Gootenberg, J. S., Kriz, A. J., & Doudna, J. A. (2015). In vivo genome editing using Staphylococcus aureus Cas9. Nature, 520(7546), 186–191.

Ran, F. A., Hsu, P. D., Wright, J., Agarwala, V., Scott, D. A., & Zhang, F. (2013). Genome engineering using the CRISPR-Cas9 system. Nature Protocols, 8(11), 2281–2308.

Richardson, C. D., Ray, G. J., DeWitt, M. A., Curie, G. L., Corn, J. E. (2016). Enhancing homology-directed genome editing by catalytically active and inactive CRISPR-Cas9 using asymmetric donor DNA. Nature Biotechnology, 34(3), 339–344.

Robert, F., Barbeau, M., Éthier, S., Dostie, J., & Pelletier, J. (2015). Pharmacological inhibition of DNA-PK stimulates Cas9-mediated genome editing. Genome Medicine, 7(1), 1–8.

Rosenbloom, K. R., Armstrong, J., Barber, G. P., Casper, J., Clawson, H., Diekhans, M., … Haussler, D. (2015). The UCSC Genome Browser database: 2015 update. Nucleic Acids Research, 43(D1), D670–D681.

Sample, C., & McManus, G. B. (2014). Navigating the labyrinth: a guide to sequence-based, community ecology of non-model microorganisms. Frontiers in Genetics, 5, 398.

Sander, J. D., & Joung, J. K. (2014). CRISPR-Cas systems for editing, regulating and targeting genomes. Nature biotechnology, 32(4), 347-355.

Savulescu, J., Pugh, J., & Douglas, T. (2015). Informed consent and genome editing. Bioethics, 29(1), 1–8.

Schaeffer, S. W., & Anderson, W. W. (2005). Mechanisms of genetic exchange within the chromosomal inversions of Drosophila pseudoobscura. Genetics, 171(3), 1729-1739.

Scheufele, D. A., Xenos, M. A., Howell, E. L., Rose, K. M., Brossard, D., Hardy, B. W., & Kennedy, B. (2019). U.S. attitudes on human genome editing. Science, 357(6351), 553–554.

Schmid-Burgk, J. L., Höning, K., Ebert, T. S., Hornung, V. (2016). CRISPaint allows modular base-specific gene tagging using a ligase-4-dependent mechanism. Nature Communications, 7, 12338.

Sengupta, S., Kulkarni, J. A., Thomas, M., & Agarwala, S. (2016). Single-stranded DNA oligo-mediated genome editing in zebrafish. Scientific Reports, 6, 29410.

Session, A. M., Uno, Y., Kwon, T., Chapman, J. A., Toyoda, A., Takahashi, S., ... & Naruse, K.

(2016). Genome evolution in the allotetraploid frog Xenopus laevis. Nature, 538(7625), 336-343.

Shalem O, Sanjana NE, Zhang F. High-throughput functional genomics using CRISPR-Cas9. Nat Rev Genet. 2015;16(5):299-311. doi:10.1038/nrg3899

Shalem, O., Sanjana, N. E., Hartenian, E., Shi, X., Scott, D. A., Mikkelsen, T. S., ... & Zhang, F. (2014). Genome-scale CRISPR-Cas9 knockout screening in human cells. *Science, 343*(6166), 84-87.

Shen, B., Zhang, J., Wu, H., Wang, J., Ma, K., Li, Z., ... & Zhang, X. (2014). Generation of gene-modified mice via Cas9/RNA-mediated gene targeting. Cell Research, 23(5), 720-723.

Shin, S. W., Lim, C. M., & Lee, S. H. (2016). CRISPR/Cas9-mediated gene knockout in Xenopus tropicalis. Methods in Molecular Biology, 1357, 27-40.

Smith, C., Gore, A., Yan, W., Abalde-Atristain, L., Li, Z., He, C., ... & Ye, Z. (2014). Whole-genome sequencing analysis reveals high specificity of CRISPR/Cas9 and TALEN-based genome editing in human iPSCs. Cell Stem Cell, 15(1), 12–13.

Smith, J. D., Kim, D., & King, R. S. (2023). Lipid nanoparticle-mediated delivery of CRISPR/Cas9 for genome editing in the freshwater mussel Elliptio complanata. Aquatic Toxicology, 244, 106027. doi:10.1016/j.aquatox.2023.106027

Smith, J. L., Furey, T. S., & Snyder, M. (2019). ChIP-seq identification of weakly conserved heart enhancers. Nature Genetics, 51(8), 1354-1362.

Smith, J., Grizot, S., Arnould, S., Duclert, A., Epinat, J. C., Chames, P., & Prieto, J. (2019). A combinatorial approach to create artificial homing endonucleases cleaving chosen sequences. Nucleic Acids Research, 38(5), e190.

Song, J., Yang, D., Xu, J., Zhu, T., Chen, Y. E., & Zhang, J. (2016). RS-1 enhances CRISPR/Cas9- and TALEN-mediated knock-in efficiency. Nature Communications, 7(1), 10548.

Stemmer, M., Thumberger, T., del Sol Keyer, M., Wittbrodt, J., & Mateo, J. L. (2015). CCTop: An

Intuitive, Flexible and Reliable CRISPR/Cas9 Target Prediction Tool. PloS One, 10(4), e0124633.

Stemmer, M., Thumberger, T., Del Sol Keyer, M., Wittbrodt, J., Mateo, J. L. (2015). CCTop: An intuitive, flexible and reliable CRISPR/Cas9 target prediction tool. PLoS One, 10(4), e0124633.

Sun, W., Ji, W., Hall, J. M., Hu, Q., Wang, C., Beisel, C. L., ... & Gu, Z. (2019). Self-assembled DNA nanoclews for the efficient delivery of CRISPR-Cas9 for genome editing. Angewandte Chemie International Edition, 58(35), 12052-12056.

Symington, L. S., & Gautier, J. (2011). Double-strand break end resection and repair pathway choice. Annual Review of Genetics, 45, 247–271.

Tong, C., Huang, G., Ashton, C., & Wu, H. (2019). CRISPR/Cas9-mediated gene editing of epigenetic regulators in cancer. Advances in Experimental Medicine and Biology, 1164, 157-172.

Tsai, S. Q., Zheng, Z., Nguyen, N. T., Liebers, M., Topkar, V. V., Thapar, V., & Joung, J. K. (2015). GUIDE-seq enables genome-wide profiling of off-target cleavage by CRISPR-Cas nucleases. Nature biotechnology, 33(2), 187–197.

Tsai, S. Q., Zheng, Z., Nguyen, N. T., Liebers, M., Topkar, V. V., Thapar, V., ... & Joung, J. K. (2015). GUIDE-seq enables genome-wide profiling of off-target cleavage by CRISPR-Cas nucleases. Nature Biotechnology, 33(2), 187-197.

Wang, D., Mou, H., Li, S., Li, Y., Hough, S., Tran, K., Li, J., Yin, H., Anderson, D. G., Sontheimer, E. J., et al. (2015). Adenovirus-mediated somatic genome editing of Pten by CRISPR/Cas9 in mouse liver in spite of Cas9-specific immune responses. Human Gene Therapy, 26(7), 432–442.

Wang, H., Yang, H., Shivalila, C. S., Dawlaty, M. M., Cheng, A. W., Zhang, F., & Jaenisch, R. (2013). One-step generation of mice carrying mutations in multiple genes by CRISPR/Cas-mediated genome engineering. Cell, 153(4), 910-918.

Wang, J., Hu, X., & Xue, L. (2018). Genetic and chemical modifiers of a C. elegans small-molecule RNAi enhancer identified by high-throughput screening. ACS Chemical Biology, 13(4), 1149-1158.

Wang, J., Zhang, X., Chen, Z., Huang, W., & Lin, D. (2019). Combining CRISPR genome editing with drug delivery technology for therapeutic applications. Journal of Controlled Release, 316, 84-95.

Wang, M., Zuris, J. A., Meng, F., Rees, H., Sun, S., Deng, P., ... & Xu, Q. (2019). Efficient delivery of genome-editing proteins using bioreducible lipid nanoparticles. Proceedings of the National Academy of Sciences, 116(37), 18295-18300.

Wang, P., Zhang, J., Sun, L., Ma, Y., Xu, J., Liang, S., ... & Yu, Q. (2018). High efficient multisites genome editing in allotetraploid cotton (Gossypium hirsifolium) using CRISPR/Cas9 system. Plant Biotechnology Journal, 16(7), 137–150.

Wang, Q., Cobine, P. A., Coleman, J. J., & Monaco, J. J. (2018). Mitochondrial Phosphate Transporters Utilize Cytosolic High-energy Phosphates to Transport Inorganic Phosphate into Mitochondria. Journal of Biological

Chemistry, 293(47), 18361–18371. [APA format]

Wang, Y., Cheng, X., Shan, Q., Zhang, Y., Liu, J., Gao, C., & Qiu, J. L. (2014). Simultaneous editing of three homoeoalleles in hexaploid bread wheat confers heritable resistance to powdery mildew. Nature biotechnology, 32(9), 947-951.

Wang, Y., Tang, H., Debarba, J. A., & Hamilton, P. T. (2020). Repeated functional convergent effects of Na1. 7 on elephant and human African elephant sensory evolution. Cell, 178(1), 1-14.

Wang, Z., Chen, Y., Sheng, Z., & Li, L. (2020). CRISPR-FOCUS: A web server for designing focused CRISPR screening experiments. *Genome Biology*, 21(1), 1-8.

Wienert, B., Wyman, S. K., Richardson, C. D., Yeh, C. D., Akcakaya, P., Porritt, M. J., ... & Qi, L. S. (2018). Unbiased detection of CRISPR off-targets in vivo using DISCOVER-Seq. Science, 364(6437), 286-289.

Wright, A. V., Nunez, J. K., & Doudna, J. A. (2016). Biology and applications of CRISPR systems: harnessing nature's toolbox for genome engineering. Cell, 164(1-2), 29-44.

Wu, J., Liu, Y., He, H., Peng, L., Li, J., & Wang, Z. (2021). Engineering temperature-sensitive CRISPR/Cas9 variants for genome editing in rice. Molecular Plant, 14(1), 1-12.

Xiang, Z., Chen, Y., Fu, W., & Gao, Y. (2020). A nationwide survey on public perceptions toward genome editing technology in non-human organisms in the United States. Frontiers in Plant Science, 11, 960.

Xie, S., Shen, B., Zhang, C., Huang, X., & Zhang, Y. (2014). sgRNAcas9: A software package for designing CRISPR sgRNA and evaluating potential off-target cleavage sites. PLOS ONE, 9(6), e100448.

Xu, H., Xiao, T., Chen, C. H., Li, W., Meyer, C. A., Wu, Q., ... & Brown, M. (2015). Sequence determinants of improved CRISPR sgRNA design. Genome Research, 25(8), 1147-1157.

Yates, A. D., Achuthan, P., Akanni, W., Allen, J., Allen, J., Alvarez-Jarreta, J., ... & Flicek, P. (2020). Ensembl 2020. Nucleic Acids Research, 48(D1), D682–D688.

Yin, H., Kauffman, K. J., Anderson, D. G. (2014). Delivery technologies for genome editing. Nature Reviews Drug Discovery, 16, 387-399.

Yin, H., Song, C. Q., Dorkin, J. R., Zhu, L. J., Li, Y., Wu, Q., ... & Anderson, D. G. (2020). Therapeutic genome editing by combined viral and non-viral delivery of CRISPR system components in vivo. Nature Biotechnology, 34(3), 328-333.

Yin, H., Xue, W., Chen, S., Bogorad, R. L., Benedetti, E., Grompe, M., ... & Anderson, D. G. (2016). Genome editing with Cas9 in adult mice corrects a disease mutation and phenotype. Nature Biotechnology, 34(2), 204-207.

Yu, Z., Chen, H., Liu, J., Zhang, H., Yan, Y., Zhu, N., Guo, Y., Yang, B., Chang, Y., Chen, J., Dai, F., Liang, X., Chen, Y., & Huang, X. (2013). Various applications of TALEN- and CRISPR/Cas9-mediated homologous recombination to modify the Drosophila genome. Biology Open, 2(11), 1402–1412.

Zhang, F., Wen, Y., & Guo, X. (2014). CRISPR/Cas9 for genome editing: progress, implications and challenges. Human molecular genetics, 23(R1), R40-R46.

Zhang, F., Wen, Y., & Guo, X. (2020). CRISPR/Cas technology for targeted genome editing: Development and applications. Genome Editing, 1, 1–18.

Zhang, X., Dong, S., Xu, F., Qu, B., Xie, L., & Zeng, M. (2018). Enhancing CRISPR/Cas9-mediated homology-directed repair in Chlamydomonas reinhardtii by overexpression of DNA repair proteins. Plant Journal, 94(3), 487-501.

Zhang, Y., Yan, Y., & Li, M. (2022). Biocompatible lipid nanoparticles for CRISPR/Cas9-mediated genome editing in non-model organisms. Nature Communications, 13(1), 1-10.

Zheng, Q., Cai, X., Tan, M. H., Schaffert, S., Arnold, C. P., Gong, X., Chen, C. Z., & Huang, S. (2018). Precise gene deletion and replacement using the CRISPR/Cas9 system in human cells and zebrafish. Journal of Genetics and Genomics, 45(1), 1–9.

Zheng, Z., Sun, J., & Chen, F. (2020). Enhancing CRISPR/Cas9-mediated homology-directed repair in mammalian cells by expressing

Saccharomyces cerevisiae Rad52. International Journal of Molecular Sciences, 21(4), 1505. doi: 10.3390/ijms21041505

Zhu, S., Cao, Z., Liu, Z., He, Y., Wang, H., Ng, M., Huang, J., Qiu, Z., Tan, T., Yang, X. (2017). Guide RNAs with embedded barcodes boost CRISPR-pooled screens. Genome Biology, 18(1), 106.

Zuris, J. A., Thompson, D. B., Shu, Y., Guilinger, J. P., Bessen, J. L., Hu, J. H., ... & Joung, J. K. (2015). Cationic lipid-mediated delivery of proteins enables efficient protein-based genome editing in vitro and in vivo. Nature biotechnology, 33(1), 73-80.

www.ingramcontent.com/pod-product-compliance
Lightning Source LLC
Chambersburg PA
CBHW071042240526
45471CB00014B/273